T0350927

Processing of Nanoparticle Materials and Nanostructured Films

Processing of Nanoparticle Materials and Nanostructured Films

Ceramic Transactions, Volume 223

Edited by
Kathy Lu
Chris Li
Eugene Medvedovski
Eugene A. Olevsky

A John Wiley & Sons, Inc., Publication

Published by John Wiley & Sons, Inc., Hoboken, New Jersey.
Published simultaneously in Canada.

Library of Congress Cataloging-in-Publication Data is available.

ISBN 978-0-470-92731-1

Printed in the United States of America.

10 9 8 7 6 5 4 3 2 1

Contents

Preface

There have been extraordinary developments in nanomaterials in the past two decades. Nanomaterial processing is one of the key components for this success. This volume, titled Processing of Nanoparticle Materials and Nanostructured Films, is a collection of the papers presented at Controlled Processing of Nanoparticle-based Materials and Nanostructured Films symposium held during the Materials Science and Technology 2009 conference (MS&T'09), October 25–29, 2009 in Pittsburgh, PA. It summarizes the progress that has been achieved most recently in understanding and processing nanoparticle-based materials and nanostructured films.

Nanoparticle-based materials and nanostructured films hold great promise to enable a broad range of new applications. This includes high energy conversion efficiency fuel cells, smart materials, high performance sensors, and structural materials under extreme environments. However, many barriers still exist in understanding and controlling the processing of nanoparticle-based materials and nanostructured films. In particular, agglomeration must be controlled in powder synthesis and processing to enable the fabrication of homogeneous green or composite microstructures, and microstructure evolution must be controlled to preserve the size and properties of the nanostructures in the finished materials. Also, novel nanostructure designs are highly needed at all stages of bulk and thin film nanomaterial formation process to enable unique performances, low cost, and green engineering. This volume focuses on three general topics, 1) Processing to preserve and improve nanoscale size, structure, and properties, 2) Novel design and understanding of new nanomaterials, such as new synthesis approaches, templating, and 3D assembly technologies, and 3) Applications of nanoparticle assemblies and composites and thin films.

We would like to thank all symposium participants and session chairs for making this symposium one of the best-attended symposia at the MS&T'09 conference. Special thanks also go out to the reviewers who devoted time reviewing the papers included in this volume. The continuous support from The American Ceramic Society is also gratefully acknowledged. This volume reflects the quality, the scope, and the quality of the presentations given and the science described during the conference.

KATHY LU
CHRIS LI
EUGENE MEDVEDOVSKI
EUGENE A. OLEVSKY

GRAPHENE ENCAPSULATED GOLD NANOPARTICLES AND THEIR CHARACTERIZATION

Junchi Wu, Nitin Chopra*
Department of Metallurgical and Materials Engineering,
Center for Materials for Information Technology (MINT)
University of Alabama
Tuscaloosa, AL-35487, USA
*Email: nchopra@eng.ua.edu
Tel: 205-348-4153
Fax: 205-348-2164

ABSTRACT

Applications of gold nanoparticles under stringent conditions (e.g. low pH) are limited by improper surface passivation and inherent instability or tendency to aggregate. These limitations can be overcome by encapsulating a gold nanoparticle inside a strong and multi-functional cage or a shell. Here, a robust batch-production method to synthesize graphene encapsulated gold nanoparticles is developed. Patterned gold nanoparticles on a silicon substrate were utilized as catalysts for the growth of graphene shells in a hydrocarbon-based chemical vapor deposition (CVD) process. The gold nanoparticles and graphene encapsulated gold nanoparticles are characterized by scanning electron microscopy (SEM) and transmission electron microscopy (TEM). As a first step towards understanding the growth mechanism of graphene shells, XPS studies were performed in order to observe the formation of gold oxide shell on gold nanoparticles. Further oxidation of the graphene shells resulted in carboxylic (-COOH) functionalities that can be very useful for developing multi-functional sensors. Finally, the post-fabrication process involving plasma oxidation/etching of graphene shells resulted in reduction of the graphene shell thickness as well as the damage to the shell by the oxidation process.

INTRODUCTION

There have been reports on the synthesis of polymer or organic molecule encapsulated nanoparticles to prevent aggregation of nanoparticles.[1,2] In most of these cases, the nanoparticle is difficult to stabilize and is incompletely protected.[3] Carbon-based encapsulation of nanoparticles, especially after the discovery of fullerenes,[4] is attracting significant interests due to the unique properties of carbon[5,6,7] and the encapsulated nanoparticle.[8,9,10,11] The carbon shell can also be a robust protective layer for the encapsulated nanoparticle.[9,12,13,14] Such nanoparticles have been prepared by arc, laser, electron irradiation, and chemical methods.[15,16,17,18] It has been observed that transition metal nanoparticles not only catalyze graphene shell growth but also carbon nanotubes (CNTs) and amorphous carbon.[19,20,21,22] In addition, for many envisioned applications, it is important that the encapsulated nanoparticle should not react with the graphene/carbon shell to form carbides.[23] Thus, formation of stable and multi-functional carbon encapsulated nanoparticles requires two major criteria that have not yet been addressed properly. First, the nanoparticle material should be both chemically inert and act as a catalyst/selective substrate for the growth of graphene or carbon shell. Second, the synthesis of such encapsulated nanoparticles should be economical even for large yields, with minimal aggregation. Critical to the graphene shell thickness, crystallinity (or extent of graphitization), and the yield, is the type of the catalyst nanoparticles.[22] To address these issues and the above listed criteria, encapsulating gold nanoparticles with graphene is very promising. Size-dependent properties of gold nanoparticles make them useful for many applications such as optical devices and chemical sensors.[24] Fabrication of a continuous and non-porous graphene shell represents a unique way to stabilize these nanoparticles and provide a robust surface passivation. Additionally, rich surface chemistry of graphene can also be utilized making such nanoparticles multi-functional and useful for novel chemical and biological sensors, analytical platforms for high throughput screening, and advanced spectroscopic tools.[20]

EXPERIMENTAL

Si wafers (B doped, <100>, International Wafer Service, Colfax, CA) were cleaned using acetone in an ultrasonication process for 25 min. Gold nanoparticles (10 nm colloidal gold, 0.01% as $HAuCl_4$, Aldrich) were dispersed on dried Si wafers by immersing the wafers in gold nanoparticle solution for 48 h. After the dispersion process, the substrates were dried and stored in a vacuum desiccator at room temperature. The substrates with dispersed gold nanoparticles were plasma oxidized using a plasma etcher in an oxygen environment. The oxidation was done in a two-step exposure to O_2-plasma, where each step is 15 min. The experiment is stopped in between the steps for 15 min to lower the temperature of the plasma chamber. The gas pressure and the forward power of the plasma system were maintained at 1200 mTorr and 170 W. The oxidized gold nanoparticles were characterized in Field Emission SEM, JEOL 7000 and X-ray Photoelectron Spectroscopy (XPS, Kratos Axis 165 XPS) to observe the presence of gold oxide. For XPS characterization, the pass energy and the current were set at 160 eV and 15 mA, respectively. The curves have been smoothed by "Quadratic Savitzky-Golay" method (kernel width of 1 and 0.5 is used).

On the other hand, similarly oxidized gold nanoparticles on a cleaned Si substrate were inserted in a CVD chamber in the presence of xylene/hydrogen/argon.[19,25,26] This resulted in the production of graphene encapsulated gold nanoparticles uniformly dispersed on a Si substrate. The dispersed graphene encapsulated gold nanoparticles were characterized by TEM. They were further plasma oxidized to observe the effects of plasma damage to the graphene shell.[25]

RESULTS AND DISCUSSION

A series of characterization and growth steps were employed to fabricate graphene encapsulated gold nanoparticles as well as to understand the oxidation of gold nanoparticles. The latter is the first step in understanding the growth mechanism of graphene encapsulated gold nanoparticles. These hybrid nanoparticles can be multi-functional, versatile platforms for fabricating advanced sensors and devices.[25] The gold nanoparticles (~10 nm diameter) were successfully dispersed on a silicon wafer as indicated in figure 1.

Successful oxidation in an oxygen plasma process formed a very thin gold oxide shell on the surface of the gold nanoparticle. Although this shell was not observed in TEM but the XPS showed a clear evidence of formation of gold oxide (Figure 2). The surface of oxidized and non-oxidized gold nanoparticle samples were examined by XPS and the comparison of the spectra indicate the formation of gold oxide. Pure element gold peaks are visible at Au-Au[1] (Au $4f_{7/2}$):

84.2 eV and Au-Au[2] (Au $4f_{5/2}$): 87.6 eV. While extra peaks are developed in oxidized samples corresponding to Au-O[1]: 85.0 eV and Au-O[2]: 88.6 eV confirming that gold oxide shell is present on these nanoparticles. These peaks correspond to the chemical shift of Au-4f levels that have been strong indicators of Au_2O_3[28] in case of surface oxidized gold nanoparticles. The heat of formation for Au_2O_3 is reported to be +19.3 kJ/mol.[27]

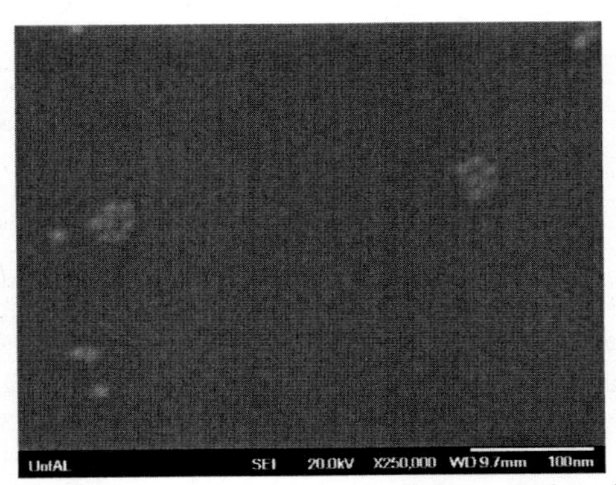

Figure 1. SEM image of gold nanoparticles dispersed on silicon substrates.

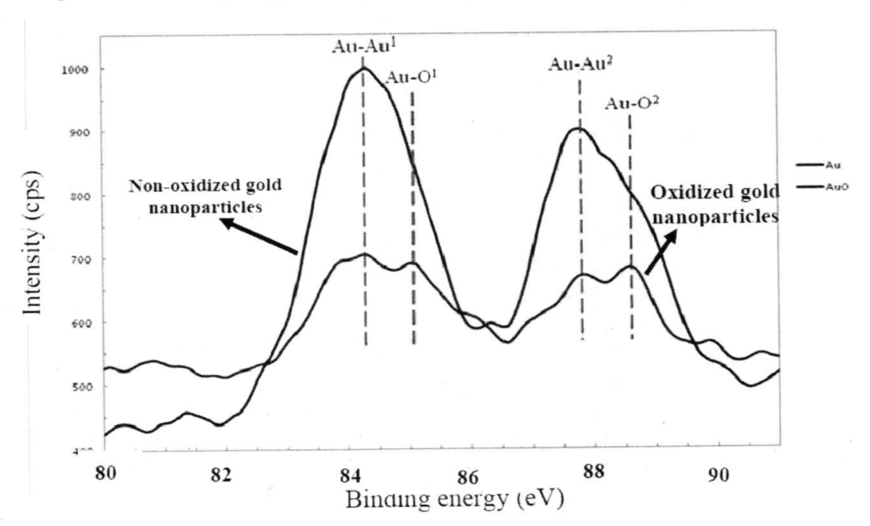

Figure 2. XPS spectra for non-oxidized (black) gold nanoparticles and oxidized (blue) gold nanoparticles. Also indicated by arrows.

Similarly oxidized gold nanoparticles were further utilized as a catalyst for the growth of graphene shells in a CVD process.[25] In our previous study, we have demonstrated that non-oxidized gold nanoparticles do not result in any graphene shells.[25] Thus, it is proposed here that the formation of graphene shell over the surface oxidized gold nanoparticles involves electron transfer process. Gold oxide being unstable at high temperatures[28], tends to reduce to gold (0) during the CVD process.[28] The electrons needed for this process are provided by the incoming carbon feed, which in turn forms sp^2 hybridized graphene shell. In this regard, our XPS studies confirm that gold oxide is present on gold nanoparticles after plasma oxidation process.

Once the graphene shell is formed on the gold nanoparticles, it imparts multi-functionality to the nanoparticles. This can be attributed to rich graphene/carbon chemistry instead of thiol-based gold chemistry that severely limits applications of gold nanoparticles. As a next step, the graphene shell can be further oxidized using plasma oxidation process and results in –COOH derivatization of graphene encapsulated gold nanoparticles.[25] These –COOH groups are readily available for further chemical functionalization and form the basis of developing novel bioanalytical systems based on graphene encapsulated gold nanoparticles. However, it is possible to damage the graphene shells if a prolonged plasma oxidation of shells is performed. In addition, this process can also result in reduction in number of graphene shells encapsulating gold nanoparticles. Thus, in this study, we also report a unique method to etch graphene shells or reduce the number of layers of the graphene shells in the graphene encapsulated gold nanoparticles. This is achieved by performing post-growth plasma etching process. As indicated in the figure 3, it was observed that the graphene shells were etched, damaged, and the number of shells decreased after performing the plasma oxidation. This is a promising approach to control the physical characteristics of these nanoparticles and is an easy handle to tune their properties as well as chemical functionalities. More detailed microscopic analysis of graphene shell is reported recently.[25]

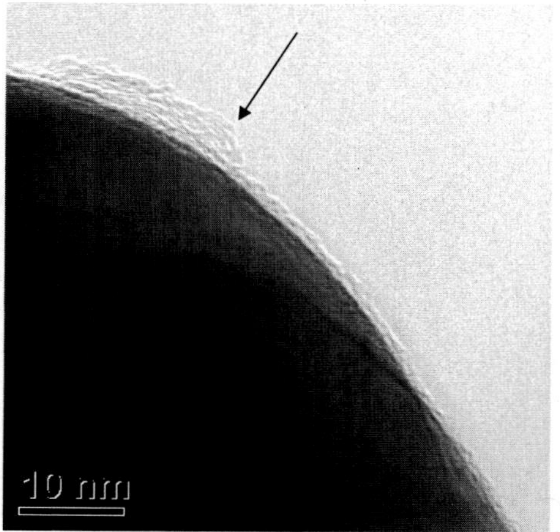

Figure 3. TEM image of a plasma oxidized/etched graphene (marked by an arrow) encapsulated gold nanoparticles. Plasma etching of graphene shells in a graphene encapsulated gold nanoparticles. The shells are damaged, etched, and removed during the process.

CONCLUSIONS

Graphene encapsulated gold nanoparticles were fabricated and dispersed on a silicon wafer. SEM, TEM, and XPS studies confirmed the chemical composition and physical characteristics of nanoparticles at various stages of the fabrication process. XPS studies reveal the formation of gold oxide after plasma oxidation of gold nanoparticles. A growth mechanism based on electron transfer process during the CVD process is proposed that facilitates the growth of graphene shell on the surface oxidized gold nanoparticles. Finally, post-CVD plasma etching removes and damages graphene shell. This dry processing or etching of graphene shell allows for controlling the characteristics of graphene encapsulated gold nanoparticles. Such multi-component and multi-functional nanoparticles are of immense importance for developing analytical devices and sensors, self-assembly platforms, and optical systems.

ACKNOWLEDGEMENTS

This material is based upon work supported by the National Science Foundation (NSF) under Grant No. (0925445, PCAN-ECCS). The authors also thank the University of Alabama (MTE department and the Office of sponsored programs), MSE PhD program, and the MINT Center for the start-up funds and for financial aid for J.W. The authors also thank the Central Analytical Facility (CAF) for electron microscopy equipment (NSF-MRI funded) and the financial support covering the instrument time, the MINT Center for providing infrastructure support such as clean room facility and various equipments, and Mr. Rich Martens, Mr. Johnny Goodwin, and Mr. Rob Holler for providing training on microscopy and surface analysis instruments. In addition, the authors also thank the University of Kentucky for providing instrument access from time to time.

REFERENCES

[1] H. Y. Chen, S. Abraham, J. Mendenhall, S. C. Delamarre, K. Smith, I. Kim, C. A. Batt, Encapsulation of Single Small Gold Nanoparticles by Diblock Copolymers, *ChemPhysChem*, **9**, 388 – 92 (2008).
[2] X. Sun, Y. Li, Colloidal Carbon Spheres and Their Core/Shell Structures with Noble-Metal Nanoparticles, *Angew. Chemie Int. Ed.*, **43**, 597 –601 (2004).
[3] J. B. Schlenoff, M. Li, H. Ly, Stability and Self-Exchange in Alkanethiol Monolayers, *J. Am. Chem. Soc.*,**117**, 12528-36 (1995).
[4] H. W. Kroto, J. R. Heath, S. C. O'Brien, R. F. Curl, R. E. Smalley, C60: Buckminsterfullerene, *Nature*, **318**, 162-63 (1985).
[5] Z. Liu, M. Winters, M. Holodniy, H. Dai, Supramolecular Chemistry on Water-Soluble Carbon Nanotubes for Drug Loading and Delivery, *Angew. Chemie, Int. Ed.*, **46**, 2023-27 (2007).
[6] B. J. Landia, S. L. Castro, H. J. Rufa, C. M. Evans, S. G. Bailey, R. P. Raffaelle, CdSe Quantum Dot-Single Wall Carbon Nanotube Complexes for Polymeric Solar Cells, *Sol. Energ. Mat. Sol. C.*, **87**, 733–746 (2005).
[7] S. Ikeda, S. Ishino, T. Harada, N. Okamoto, T. Sakata, H. Mori, S. Kuwabata, T. Torimoto, M. Matsumura, Ligand-Free Platinum Nanoparticles Encapsulated in a Hollow Porous Carbon Shell

as a Highly Active Heterogeneous Hydrogenation Catalyst, *Angew. Chemie Int. Ed.*, **45**, 7063 – 66 (2006).
[8] M. Kim, K. Sohn, H. B. Na, T. Hyeon, Synthesis of Nanorattles Composed of Gold Nanoparticles Encapsulated in Mesoporous Carbon and Polymer Shells, *Nano Lett.*, **2**, 1383-87 (2002).
[9] J. Geng, D. A. Jefferson, F. G. Johnson, Direct Conversion of Iron Stearate into Magnetic Fe and Fe3C Nanocrystals Encapsulated in Polyhedral Graphite Cages, *Chem. Comm.*, 2442-43 (2004).
[10] E. Sutter, P. Sutter, R. Calarco, T. Stoica, R. Meijers, Assembly of Ordered Carbon Shells on GaN Nanowires, *App. Phys. Lett.*, **90**, 093118/1-3 (2007).
[11] E. Sutter, P. Sutter, Au-Induced Encapsulation of Ge Nanowires in Protective C Shells, *Adv. Mater.*, **18**, 2583-88 (2006).
[12] Y. Ma, Z. Hu, L. Yu, Y. Hu, B. Yue, X. Wang, Y. Chen, Y. Lu, Y. Liu, J. Hu, Chemical Functionalization of Magnetic Carbon-Encapsulated Nanoparticles Based on Acid Oxidation, *J. Phys. Chem. B*, **110**, 20118-122 (2006).
[13] E. Flahaut, F. Agnoli, J. Sloan, C. O'Connor, M. L. H. Green, CCVD Synthesis and Characterization of Cobalt-Encapsulated Nanoparticles, *Chem. Mater.*, **14**, 2553-58 (2002).
[14] E. Sutter, P. Sutter, Au-Induced Encapsulation of Ge Nanowires in Protective C Shells, *Adv. Mater.*, **18**, 2583-88 (2006).
[15] Y. Saito, Nanoparticles and Filled Nanoscapsule, *Carbon*, **33**, 979-88 (1995).
[16] V. P. Dravid, J. J. Host, M. H. Teng, B. Elliott, J. Hwang, D. L. Johnson, T. O. Mason, J. R. Weertman, Controlled Size Nanocapsule, *Nature*, **374**, 602 (1995).
[17] J. R. Heath, S. C. O'Brien, Q. Zhang, Y. Liu, R. F. Curl, H. W. Kroto, F. K. Tittel, R. E. Smalley, Lanthanum Complexes of Spheroidal Carbon Shells, *J. Am. Chem Soc.*, **107**, 7779-80 (1985).
[18] S. C. Tsang, Y. K. Chen, P. J. F. Harris, M . L. H. Green, A Simple Chemical Method for Opening and Filling Carbon Nanotubes, *Nature*, **372**, 159-62 (1994).
[19] N. Chopra, P. D. Kichambare, R. Andrews, B. J. Hinds, Control of Multiwalled Carbon Nanotube Diameter by Selective Growth on the Exposed Edge of a Thin Film Multilayer Structure, *Nano Lett.*, **2**, 1177-81 (2002).
[20] N. Chopra, V. G. Gavalas, B. J. Hinds, L. G. Bachas, Functional One-Dimensional Nanomaterials: Applications in Nanoscale Biosensors, *Anal. Lett.*, **40**, 2067-96 (2007).
[21] M. Nagatsu, T. Yoshida, M. Mesko, A. Ogino, T. Matsuda, T. Tanaka, H. Tatsuoka, K. Murakami, Narrow Multi-Walled Carbon Nanotubes Produced by Chemical Vapor Deposition using Graphene Layer Encapsulated Catalytic Metal Particles, *Carbon*, **44**, 3336–41 (2006).
[22] M. Meyyappan, *Carbon Nanotubes: Science and Applications*, CRC Press, Boca Raton, FL, (2005)
[23] R. S. Ruoff, D. C. Lorents, B. Chan, R. Malhotra, S. Subramoney, Single Crystal Metals Encapsulated in Carbon Nanoparticles, *Science*, **259**, 346-48 (1993).
[24] M. –C. Daniel, D. Astruc, Gold Nanoparticles: Assembly, Supramolecular Chemistry, Quantum-Size-Related Properties, and Applications toward Biology, Catalysis, and Nanotechnology, *Chem. Rev.*, **104**, 293-346 (2004).
[25] N. Chopra, L. G. Bachas, M. Knecht, Fabrication and Biofunctionalization of Carbon-Encapsulated Au Nanoparticles, *Chem. Mater.*, **21**, 1176-78 (2009).

[26] R. Andrews,D. Jacques, A. M. Rao, F. Derbyshire, D. Qian, X. Fan, E. C. Dickey, J. Chen, Continuous production of aligned carbon nanotubes: a step closer to commercial realization, *Chem. Phys. Lett.*, ***303***, 467-74 (1999).
[27] A. Krozer, M. Rodhani, M. X-ray Photoemission Spectroscopy Study of UV/Ozone Oxidation of Au Under Ultrahigh Vacuum Conditions, J. Vac. Sci. Tech. A, **15**, 1704-09 (1997).
[28] H. Tsai, E. Hu, K. Perng, M. Chen, J.-C. Wu, Y.-S. Chang, Instability of Gold Oxide Au_2O_3, Surf. Sci., **537**, L447-L450 (2003).

A SURFACTANT-ASSISTED SOLID-STATE SYNTHESIS OF $BaTiO_3$ FROM $BaCO_3$ AND TiO_2

Yu-Lun Chang, and Hsing-I Hsiang*
Department of Resources Engineering, National Cheng Kung University
Tainan, Taiwan, R.O.C.

ABSTRACT

Barium titanate powder plays an important role in versatile applications of ferroelectric materials. However, the barium titanate nanosized powder is difficult to be obtained through a conventional high-temperature solid-sate reaction. In this study, the barium titanate nanosized powder was synthesized by the solid-state reaction with surfactant assistance. The experimental results indicate that the mixing homogeneity between $BaCO_3$ and TiO_2 can be improved by the addition of polyethyleneimine (PEI), a cationic polyelectrolyte, as a surfactant. The intimate contact between $BaCO_3$ and TiO_2 particles was suggested to enhance the interfacial reaction and the subsequent diffusion process. In addition, using $BaCO_3$ with small size can further facilitate the solid-state synthesis at lower temperatures. As a result, nearly pure $BaTiO_3$ product with particle sizes less than 50nm was prepared at 800°C for 4h.

INTRODUCTION

Because of the superior dielectric and ferroelectric property, barium titanate ($BaTiO_3$) powder is widely used in the multi-layer ceramic capacitors (MLCC's). With a growth in demand, the trend of MLCC's is progressive toward smaller size and higher capacity. Thinner (below 1μm) green sheets are imperatively required, and the fabrication technique of fine and pure $BaTiO_3$ powders becomes more important. Generally, the fabrications of $BaTiO_3$ powders can be broadly divided into two categories, i.e. the solid-state reaction, and the liquid-phase reaction. Fine and pure $BaTiO_3$ powders can easily be obtained by the liquid-phase process, such as sol-gel route[1], oxalate precipitation[2], Pechini method[3] and hydrothermal method[4]. However, the high manufacturing costs and the low productivity do not meet the demand for mass-production.

Most $BaTiO_3$ powder is conventionally synthesized by a cheap and convenient solid-phase process, in which $BaCO_3$ and TiO_2 are often used as the starting materials. Nevertheless, the reaction between $BaCO_3$ and TiO_2 is usually conducted at a very high temperature, that easily causes coarsening and agglomeration of the resultant particles. Recently, the low-temperature solid-state reaction (< 1000°C) to form fine $BaTiO_3$ powders is developed numerously.[5-17] Buscaglia et al. synthesized $BaTiO_3$ using nanosized particles as the raw materials. Their results indicate that an increase of the contact points number between the reactants can facilitate the formation of $BaTiO_3$. Pure and fine $BaTiO_3$ powders can be obtained successfully at 700-800°C.[5] Ando et al. synthesized nanosized $BaTiO_3$ by a mechanical pretreatment of starting materials. The treated mixtures with high mixing homogeneity showed rather low calcination temperature to form $BaTiO_3$. Moreover the diffusion Ba^{2+} along the surfaces of resultant $BaTiO_3$ nanosized particles was supposed to take place at the early reaction stage. A pure $BaTiO_3$ with sizes <200nm was obtained at a temperature as low as 900°C.[6]

The formation mechanism of $BaTiO_3$ in the solid-state reaction has been studied extensively.[18-20] The synthesis of $BaTiO_3$ by a solid-state reaction between $BaCO_3$ and TiO_2 can be assigned to the multi-step reactions. At the beginning, $BaTiO_3$ is suggested to be formed at the contact points between $BaCO_3$ and TiO_2 at the initial stage. Then, the formation of Ba_2TiO_4 occurs at the interface between $BaTiO_3$ and $BaCO_3$. At the final stage, the Ba_2TiO_4 reacts with TiO_2 to form pure $BaTiO_3$ at the terminal stage. The formation of $BaTiO_3$ is believed to be dominated by the diffusion of barium species throughout the perovskite layer. Buscaglia et al. ever reported the kinetic results of the reaction between $BaCO_3$ and TiO_2.[5] It indicates that the data at lower temperatures is reasonably comparable to the diminishing-core reaction described by Valensi-Cater equation. However, at higher temperatures of 800°C, the data can not be satisfactorily described using any diminishing-core models. This is probably attributed to the complexity of the reaction mechanism which cannot be reduced to a simple case of phase-boundary-controlled growth or diffusion controlled growth over the entire temperature range. The conflicting statements and inadequate supports show that the solid-state reaction mechanism is still not well understood.

In this study, the low-temperature solid-state reaction to form fine $BaTiO_3$ powders is conducted from $BaCO_3$ and TiO_2 with the addition of polyethyleneimine (PEI) as a surfactant. PEI is a cationic polyelectrolyte which is widely used as adhesives, dispersion stabilizers, thickeners, and flocculating agents. It consists of primary, secondary, and tertiary amine groups with 1:2:1 ratio.[21] The protonation of amine groups can result in the positive charged surface of PEI, thereby providing an electrostatic force for adsorption on the inorganic materials.[22,23] It is expected to improve the mixing homogeneity between the reactants that facilitates the formation of $BaTiO_3$ at lower temperatures. In addition, the formation mechanism of $BaTiO_3$ in the solid-state synthesis with PEI addition was carefully examined by XRD, DTA/TG, TEM and zeta potential.

EXPERIMENTAL

The properties of $BaCO_3$ and TiO_2 powders used in this study are shown in Table 1, which are sponsored by YAGEO Corporation. For the study of the reaction mechanism, the mixtures were directly obtained by magnetically stirring mixing of the raw reactants, i.e. $BaCO_3$ (BN1) and TiO_2 (YT10) in the aqueous solution at pH = 8.4 for 12 h. Besides, the mixtures with different amount of polyethyleneimine (PEI) (Alfa Aesar, M.W. =10000 g/mol, 99%) addition were prepared with pretreatment of $BaCO_3$ (BN1 or BN3), i.e. 5g $BaCO_3$ powder was first added in 100ml PEI aqueous solution (i.e. 0.5mg/ml, 1.0mg/ml, and 1.5mg/ml) at pH = 10.5 and under magnetically stirring for 18 h. Then, the modified $BaCO_3$ and TiO_2 (YT10) were also mixed by magnetically stirring in the aqueous solution at pH = 8.4 for 12 h. The obtained dried mixtures were then calcined at different temperatures.

The thermal treatment condition was conducted at a heating rate of 10°C/min.

The crystalline phase identification was determined using X-ray diffractometry (Siemens, D5000) with Cu-K_α radiation. Electrophoretic measurements of the starting materials under different pH values were performed on a zeta potentiometer (Malvern, Zetasizer, Nano ZS). The TEM (Jeol, JEM-3010) was used to observe the crystallite size and morphology. The diffraction patterns of the crystalline species were also obtained using TEM with a camera constant of 80 cm. Semi-quantitative determination of the element content was detected using EDS (Noran, Voyager 1000,) attached to the TEM. The DTA/TG analysis was performed using a thermal analysis instrument (Netzsch STA, 409 PC) under 40ml/min air flow rate. The intensities of electron probe X-ray microanalyzer (JEOL, 8900R) were used to determine the mixing homogeneity of reactants. The variation coefficient was represented as σ/x, where σ and x denote the standard deviation and the average value, respectively.

RESULTS AND DISCUSSION

A. EFFECT OF PEI ADDITION

In the aqueous system, $BaCO_3$ and TiO_2 have isoelectric points (I.E.P.) at pH = 7.7 and pH = 7, respectively. While PEI is added, the adsorption of PEI on the $BaCO_3$ leads to highly positive zeta potential values (about 20-30 mV) of $BaCO_3$ in the pH range of 7 - 9.5 (Fig. 1). The surface positive potential is supposed to prevent the agglomeration of $BaCO_3$. Besides, the oppositely charged surfaces between PEI-added $BaCO_3$ and TiO_2 are also possible to cause an electrostatic attractive force between them to form hetero-coagulations that improve the mixing homogeneity. Figure 2 shows EPMA microscopy of the mixture of TiO_2 and $BaCO_3$ without and with 1.0 mg/ml PEI addition at pH = 8.4. The PEI added mixture reveals more even distribution of both Ba and Ti elements than the mixture without PEI. The variation coefficients of Ba distribution and Ti distribution of the mixture with PEI are 0.07 and 0.05, respectively. For the mixture without PEI, those of Ba distribution and Ti distribution are 0.11 and 0.07, respectively. It indicates that the PEI addition can improve the mixing homogeneity between $BaCO_3$ and TiO_2.

Figure 3 shows derivative thermogravimetry (DTG) curves of the mixture of $BaCO_3$ (BN1) and TiO_2 (YT10) with different amount of PEI surfactant. The mixture of $BaCO_3$ and TiO_2 without PEI roughly has two stages of weight loss. The first stage with a very small amount of reaction occurs at around 570°C, and the second reaction takes place at a temperature around 720-980°C. Besides, there is a discontinuous change along the curve at around 800°C. It may be the Hedvall effect resulted from the polymorphic transformation of $BaCO_3$.[24] Dissimilarly, the mixtures with PEI addition show three stages of weight loss. The first stages of weight loss begin at 570°C, which is as same as the mixture without PEI. The second weight losses for the mixtures with PEI occur at a specific temperature near 650°C. The first stages and the second stage of weight loss seem to be enhanced with an increase of the PEI addition. The third weight losses of the mixtures with PEI occur at temperatures above 800°C, which ends at a higher temperature than the mixture without PEI. Since the weight loss does not change significantly for the mixture with PEI addition above 1.0 mg/ml, the mixture with 1.0 mg/ml PEI was used to carry out the subsequent investigations.

Figure 4 shows XRD patterns of the calcined mixtures with 1.0 mg/ml PEI. It shows a trace amount of $BaTiO_3$ resulted in the sample at 700°C. By increasing the calcination temperature, the amount of resulted $BaTiO_3$ increases obviously. In addition, significant depletion of TiO_2 was also observed, i.e. at 600°C, the intensity of diffraction peak of TiO_2 (rutile, $2\theta = 27.5$) is grater than that of $BaCO_3$ ($2\theta = 27.7$), however, by increasing the temperature to 800°C, the diffraction intensity of TiO_2 decreases. As the temperature was up to 1000°C, a great amount of $BaTiO_3$ is formed, and the formation of by-product, Ba_2TiO_4, is also observed in the sample. At 1100°C, a sample of nearly pure $BaTiO_3$ is obtained. From the XRD observation, the first stage and second stage of weight loss in the DTG curve of mixtures with 1.0 mg/ml PEI can be attributed to the reactions for forming $BaTiO_3$. Moreover, a significant depletion of TiO_2 is also observed in the second weight loss. For the third stage of weight loss, the reaction is composed of the formation of both $BaTiO_3$ and Ba_2TiO_4.

B. EFFECT OF $BaCO_3$ SIZE

In order to investigate the effect of $BaCO_3$ size on the reaction, a small $BaCO_3$ (BN3) with 1.0 mg/ml PEI is also performed. Figure 5 shows DTG curves of the mixture using large $BaCO_3$ (BN1) and small $BaCO_3$ (BN3) with 1.0 mg/ml PEI. Compared to the mixture using large $BaCO_3$ (BN1), both significant increases of weight losses at 570°C and 650°C for the mixture using small $BaCO_3$ (BN3) are observed. Besides, the third stage of weight loss of the mixture using small $BaCO_3$ (BN3) also finishes at a lower temperature.

Figure 6 and 7 show XRD patterns of the calcined samples using large $BaCO_3$ (BN1) and small $BaCO_3$ (BN3), respectively. By increasing the holding time at 800°C, the remaining $BaCO_3$ of the sample using large $BaCO_3$ (BN1) diminish gradually. As the holding time prolongs to 8h, the remaining $BaCO_3$ almost disappears, and the $BaTiO_3$ becomes the predominated species. Besides, the formation of Ba_2TiO_4 is also observed. On the other hand, the formation of $BaTiO_3$ is facilitated in the sample using small $BaCO_3$ (BN3). It shows no more remaining $BaCO_3$ when the holding time is longer than 1h. Nearly pure $BaTiO_3$ with a very trace amount of rutile can be obtained at 800°C for holding time $\geq$ 3h. Fig. 8 shows the TEM image of the resulted $BaTiO_3$ sample at 800°C for 4h. It has cubic-like crystallites with sizes around 50 nm.

C. THE REACTION MECHANISM

Based on Ando's report[6], the weight loss of the mixture of $BaCO_3$ and TiO_2 can be separated into two stages. The first stage is related to the contact between $BaCO_3$ and TiO_2. An increase of the number of contact points of reactants can enhance the reaction in this stage. The second reaction stage is mainly attributed to the reaction between remaining $BaCO_3$ and TiO_2. These stages are similar to those proposed by Buscaglia et al..[25] In those stages, the first stage of the reaction is dominated by nucleation and growth of $BaTiO_3$ at the TiO_2-$BaCO_3$ contact points. When the residual TiO_2 are completely covered by the resulted $BaTiO_3$ product phase, the second reaction stage, which may proceeds by both surface diffusion and lattice diffusion, takes place. In this study, the DTG curve of the mixture of TiO_2 and $BaCO_3$ without PEI addition shows coincidentally the two stages of weight loss. Nevertheless, the mixture of $BaCO_3$ and TiO_2 shows three stages of weight loss when PEI is added. It reveals that the formation mechanism may consist of other reaction stage.

In the solid-state reaction between $BaCO_3$ and TiO_2, the formation of $BaTiO_3$ direct from $BaCO_3$ and TiO_2 is the thermodynamically favored reaction (Fig. 9). The decomposition of $BaCO_3$ is regarded as the catalysis of TiO_2.[24,27] Then, the nucleation of $BaTiO_3$ is preferred by the diffusion of BaO toward TiO_2[28,29], and the growth is also suggested by the diffusion of BaO through the $BaTiO_3$ product layer.[15,18,19,30-32] As the product layer increases in thickness, $BaCO_3$ and TiO_2 are separated, and the decomposition of $BaCO_3$ by the catalysis of TiO_2 no longer occurs. Then, the reaction may be followed by the diffusion of another species. Lotnyk et al. investigated the thermal diffusion between $BaCO_3$ thin film and TiO_2 single crystal, and reported that the diffusion of Ti^{4+} into the $BaCO_3$ film in the temperature range of 600-800°C.[33,34] On the other hand, Leew et al. reported that the solubility of TiO_2 within $BaTiO_3$ is approximately 2-3 at%, which is much higher than that of BaO within $BaTiO_3$.[35] Thus, the diffusion of TiO_2 through $BaTiO_3$ toward $BaCO_3$ is expected to dominate the subsequent reaction, which is also suggested to cause the significant depletion of TiO_2 in XRD result (Fig. 4). At a higher temperature, reaction between $BaCO_3$ and $BaTiO_3$ to form Ba_2TiO_4 also becomes favored thermodynamically (Fig. 9). The decomposition of $BaCO_3$ is promoted again by the catalysis of $BaTiO_3$, and the diffusion of BaO toward $BaTiO_3$ for the formation of Ba_2TiO_4 takes place. It proceeds until the weight loss levels off. Finally, the mechanism that TiO_2 diffuses through the product layer to react with Ba_2TiO_4 to form pure $BaTiO_3$ is also suggested at a higher temperature.

The mixture with PEI addition has a good mixing homogeneity between $BaCO_3$ and TiO_2, thereby facilitating the initial reaction to form $BaTiO_3$. Then, a great amount of resulted $BaTiO_3$ can provide a lot of paths for TiO_2 diffusing through. The diffusion of TiO_2 is suggested to dominate the formation of $BaTiO_3$ that contributed to significant weight losses from 650°C to 800°C. For the PEI-added mixture using small $BaCO_3$, it not only increases the contact area between the reactants, but also reduces the path length for TiO_2 diffusion. Therefore, the initial reaction between $BaCO_3$ and TiO_2 was enhanced as well, and a great amount of $BaTiO_3$ can be formed at a low temperature. In this study, nearly pure $BaTiO_3$ can be obtained at 800°C for holding time $\geq$ 3h.

CONCLUSIONS

A possible reaction mechanism for the solid-state reaction between $BaCO_3$ and TiO_2 to form $BaTiO_3$ is proposed. A multiple diffusion mechanism can be used to well explain the reaction results. The addition of PEI surfactant can modify the surface charge of $BaCO_3$ from negative into positive, causing a strong electrostatic attractive force between $BaCO_3$ and TiO_2 to form hetero-coagulations.

An increase of the number of contact point between $BaCO_3$ and TiO_2 leads to an enhancement of the initial reaction between $BaCO_3$ and TiO_2 to form $BaTiO_3$. As the product layer increases in thickness, $BaCO_3$ and TiO_2 are separated, and the decomposition of $BaCO_3$ by the catalysis of TiO_2 no longer occurs. Then the reaction may be followed by the diffusion of TiO_2. For using small $BaCO_3$, it not only increases the contact area between the reactants, but also reduces the path length for TiO_2 diffusion. Therefore, a great amount of $BaTiO_3$ can be formed at a low temperature. In this study, nearly pure $BaTiO_3$ can be obtained at 800°C for holding time $\geq$ 3h. The resulted $BaTiO_3$ sample at 800°C for 4h shows cubic-like crystallites with sizes around 50 nm.

REFERENCES

[1] U.Y. Hwang, H.S. Park, and K.K. Koo, "Low-Temperature Synthesis of Fully Crystallized Spherical $BaTiO_3$ Particles by the Gel-Sol Method," *J. Am. Ceram. Soc.*, **87**, 2168-2174 (2004).

[2] Z.H. Park, H.S. Shin, B.K. Lee, and S.H. Cho, "Particle Size Control of Barium Titanate Prepared from Barium Titanyl Oxalate," *J. Am. Ceram. Soc.*, **80**, 1599-1604 (1997).

[3] T.T. Fang, and T.D. Tsay, "Effect of pH on the Chemistry of the Barium Titanium Citrate Gel and Its Thermal Decomposition Behavior," *J. Am. Ceram. Soc.*, **84**, 2475-2478 (2001).

[4] P. Princeloup, C. Courtois, A. Leriche, and B. Thierry, "Hydrothermal Synthesis of Nanometer-Sized Barium Titanate Powders: Control of Barium/Titanium Ratio, Sintering, and Dielectric Properties," *J. Am. Ceram. Soc.*, **82**, 3049-3056 (1999).

[5] M.T. Buscaglia, M. Bassoli, and V. Buscaglia, "Solid-State Synthesis of Ultrafine $BaTiO_3$ Powders from Nanocrystalline $BaCO_3$ and TiO_2," *J. Am. Ceram. Soc.*, **88**, 2374-2379 (2005).

[6] C. Ando, R. Yanagawa, H. Chazono, H. Kishi, and M. Senna, "Nuclei-Growth Optimization for Fine-Grained $BaTiO_3$ by Precision-Controlled Mechanical Pretreatment of Starting Powder Mixture," *J. Mater. Res.*, **19**, 3592-3599 (2004).

[7] D.F.K. Hennings, B. S. Schreinemacher, and H. Schreinemacher, "Solid-State Preparation of $BaTiO_3$-Based Dielectrics, Using Ultrafine Raw Materials," *J. Am. Ceram. Soc.*, **84**, 2777-82 (2001).

[8] J. Xue, J. Wang, and D. Wan, "Nanosized Barium Titanate Powder by Mechanical Activation," *J. Am. Ceram. Soc.*, **83**, 232-34 (2000).

[9] C. Ando, H. Kishi, H. Oguchi, and M. Senna, "Effects of Bovine Serum Albumin on the Low Temperature Synthesis of Barium Titanate Microparticles via a Solid State Route," *J. Am. Ceram. Soc.*, **89**, 1709-1712 (2006).

[10] V. Berbenni, A. Marini, and G. Bruni, "Effect of Mechanical Milling on Solid State Formation of $BaTiO_3$ from $BaCO_3$-TiO_2 (Rutile) Mixtures," *Thermochim. Acta*, **374**, 151-158 (2001).

[11] C. Gomez-Yanez, C. Benitez, and H. Balmori-Bamirez, "Mechanical Activation of the Synthesis Reaction of $BaTiO_3$ from a Mixture of $BaCO_3$ and TiO_2 Powders," *Ceram. Int.*, **26**, 271-277 (2000).

[12] R. Yanagawa, M. Senna, C. Ando, H. Chazono, and H. Kishi, "Preparation of 200 nm $BaTiO_3$ Particles with their Tetragonality 1.010 via a Solid-State Reaction Preceded by Agglomeration-Free

Mechanical Activation," *J. Am. Ceram. Soc.*, **90**, 809-814 (2007).

[13]S.S. Ryu, S.K. Lee, and D.H. Yoon, "Synthesis of Fine Ca-doped $BaTiO_3$ Powders by Solid-state Reaction Method-Part I: Mechanical Activation of Starting Materials," *J. Electroceram.*, **18**, 243-50 (2007).

[14]W. Chaisan, R. Yimnirun, and S. Ananta, "Effect of Vibro-Milling Time on Phase Formation and Particle Size of Barium Titanate Nanopowders," *Ceram. Int.*, **35**, 173-176 (2009).

[15]E. Brzozowski, and M.S. Castro, "Synthesis of Barium Titanate Improved by Modifications in the Kinetics of the Solid State Reaction," *J. Eur. Ceram. Soc.*, **20**, 2347-2351 (2000).

[16]K. Kobayashi, T. Suzuki, and Y. Mizuno, "Microstructure Analysis of Solid-State Reaction in Synthesis of $BaTiO_3$ Powder Using Transmission Electron Microscope," *Appl. Phy. Exp.* **1**, 041602 (2008).

[17]M.T. Buscaglia, V. Buscaglia, and R. Alessio, "Coating of $BaCO_3$ Crystals with TiO_2: Versatile Approach to the Synthesis of $BaTiO_3$ Tetragonal Nanoparticles," *Chem. Mater.* **19**, 711-718 (2007).

[18]A. Beauger, J.C. Mutin, and J.C. Niepce, "Synthesis Reaction of Metatitanate $BaTiO_3$, Part 1 Effect of the Gaseous Atmosphere Upon the Thermal Evolution of the System $BaCO_3$-TiO_2," *J. Mater. Sci.*, **18**, 3041-46 (1983).

[19]A. Beauger, J.C. Mutin, and J.C. Niepce, "Synthesis Reaction of Metatitanate $BaTiO_3$, Part 2 Study of solid-solid reaction interfaces," *J. Mater. Sci.*, **18**, 3543-50 (1983).

[20]J.C. Niepce, and G. Thomas,"About the Mechanism of the Solid-Way Synthesis of Barium Metatitanate, Industrial Consequences," *Solid State Ionics*, **43**, 69-76 (1990).

[21]C. R. Dick, and G. E. Ham, "Characterization of Polyethylenimine," *J. Macromol Sci. Chem.*, **A4**,1301-1314 (1970).

[22]C.C. Chung, and J.H. Jean, "Aqueous Synthesis of Y_2O_2S: Eu / Silica Core-Shell Particles," *J. Am. Ceram. Soc.*, **88**, 1341-1344 (2007).

[23]X. Zhu, T. Uchikoshi, T.S. Suzuki, and Y. Sakka, "Effect of Polyethylenimine on Hydrolysis and Dispersion Properties of Aqueous Si_3N_4 Suspensions," *J. Am. Ceram. Soc.*, **90**, 797-804 (2007).

[24]J. Bera, and D. Sarkar, "Formation of $BaTiO_3$ from Barium Oxalate and TiO_2," *J. Electroceram.*, **11**, 131-137 (2003).

[25]M.T. Buscaglia, M. Bassoli, and V. Buscaglia, "Solid-State Synthesis of Nanocrystalline $BaTiO_3$: Reaction Kinetics and Powder Properties," *J. Am. Ceram. Soc.*, **91**, 2862-2869 (2008).

[26]L.K. Templeton, and J.A. Pask, "Formation of $BaTiO_3$ from $BaCO_3$ and TiO_2 in Air and in CO_2," *J. Am. Ceram. Soc.*, **42**, 212-216 (1959).

[27]T. Ishii, R. Furuichi, T. Nagasawa, and K. Yokoyama, "The Reactivities of TiO_2 (Rutile and Anatase) for the Solid-state Reactions with $BaSO_4$ and $BaCO_3$," *J. Therm. Anal. Calorim.*, **19**, 467-474 (1980).

[28]K. Kobayashi, T. Suzuki, and Y. Mizuno, "Microstructure Analysis of Solid-State Reaction in Synthesis of $BaTiO_3$ Powder Using Transmission Electron Microscope," *Appl. Phys. Exp.*, **1**, 041602 (2008).

[29]U.Manzoor, and D.K.Kim, "Synthesis of Nano-sized Barium Titanate Powder by Solid-state Reaction between Barium Carbonate and Titania," *J. Mater. Sci. Technol.*, **23**, 655-658 (2007).

[30]E. Brzozowski, J. Sanchez, and M. S. Castro, "$BaCO_3$-TiO_2 Solid State Reaction: A Kinetic Study," *J. Mater. Synth. Process.*, **10**, 1-5 (2002).

[31]E. Brzozowski, M.S. Castro, "Lowering the Synthesis Temperature of High-purity $BaTiO_3$ Powders by Modifications in the Processing Conditions," *Thermochim. Acta,* **398**, 123-129 (2003).

[32]Y.H. Hu, M.P. Harmer, and D.M. Smyth, "Solubility of BaO in $BaTiO_3$," *J. Am. Ceram. Soc.*, **68**, 372-376 (1985).

[33]A. Lotnyk, S. Senz, and D. Hesse, "Formation of $BaTiO_3$ Thin Films from (110) TiO_2 Rutile Single Crystals and $BaCO_3$ by Solid State Reactions," *Solid State Ionics*, **177**, 429-436 (2006).

[34]A. Lotnyk, S. Senz, and D. Hesse, "Thin-Film Solid-State Reactions of Solid $BaCO_3$ and BaO vapor with (100) Rutile Substrates," *Acta Mater.*, **55**, 2671-81 (2007).

[35]S. Leew, C.A. Randall, and Z.K. Liu, "Modified Phase Diagram for the Barium Oxide-Titanium Dioxide System for the Ferroelectric Barium Titanate," *J. Am. Ceram. Soc.*, **90**, 2589-2594 (2007).

Table 1. Properties of raw materials.

	Phase composition	Purity (%)	BET (m^2/g)	d_{50} (μm)
YT10	TiO_2, 91% rutile and 9% anatase	99.9	11	0.369
BN1	$BaCO_3$,	98.7	2.6	6.131
BN3	$BaCO_3$,	97.6	30.6	0.174

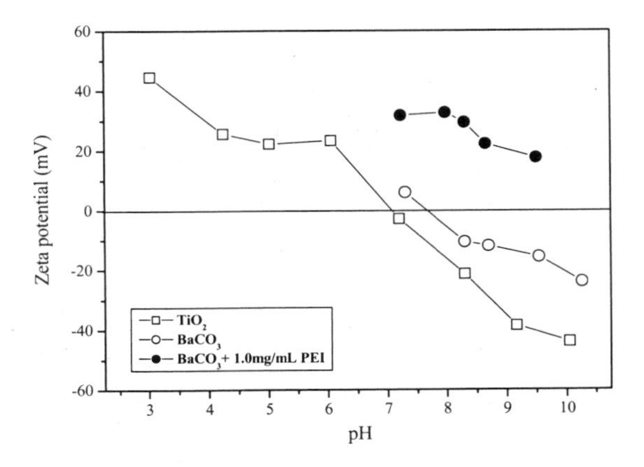

Figure 1. Zeta potential as a function of pH for the TiO_2, and $BaCO_3$ powders with and without PEI addition.

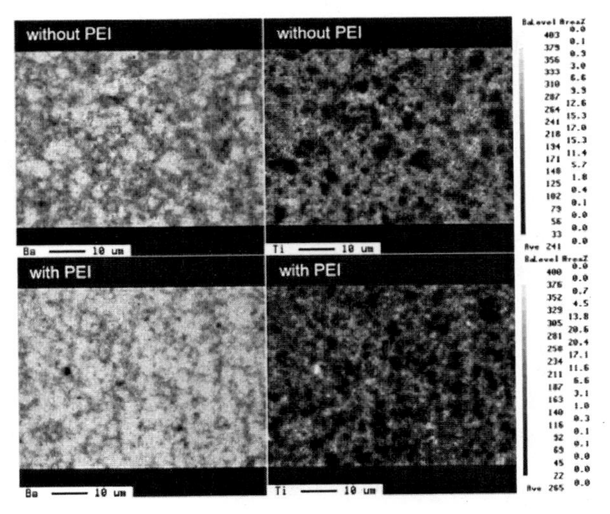

Figure 2. EPMA images of mixture of TiO_2 and $BaCO_3$ powders with and without PEI addition.

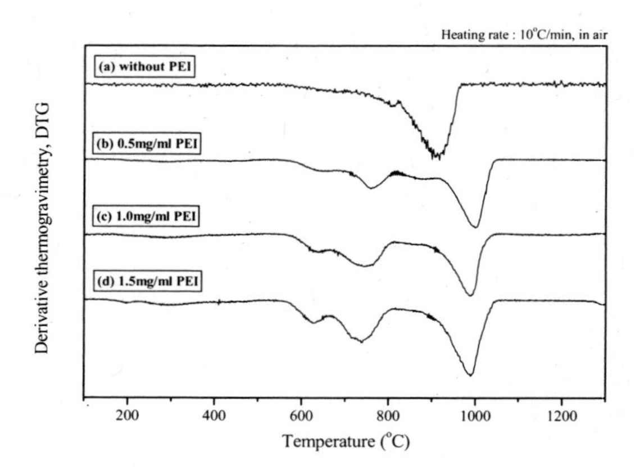

Figure 3. DTG curves of the mixture of TiO_2 and large $BaCO_3$ (BN1) with heating rate of 10°C/min, (a) without PEI, (b) with 0.5 mg/ml PEI, (c) with 1.0 mg/ml PEI, and (d) with 1.5 mg/ml PEI.

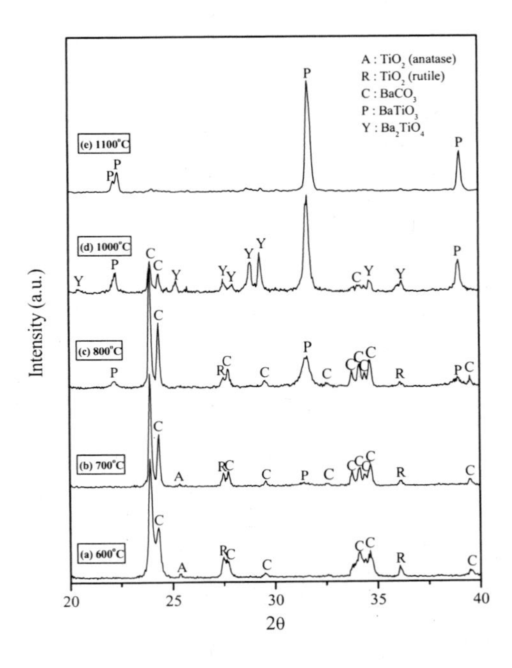

Figure 4. XRD patterns of the mixture of TiO_2 and large $BaCO_3$ (BN1) with 1.0 mg/ml PEI calcined at different temperatures, (a) 600°C, (b) 700°C, (c) 800°C, (d) 1000°C, and 1100°C.

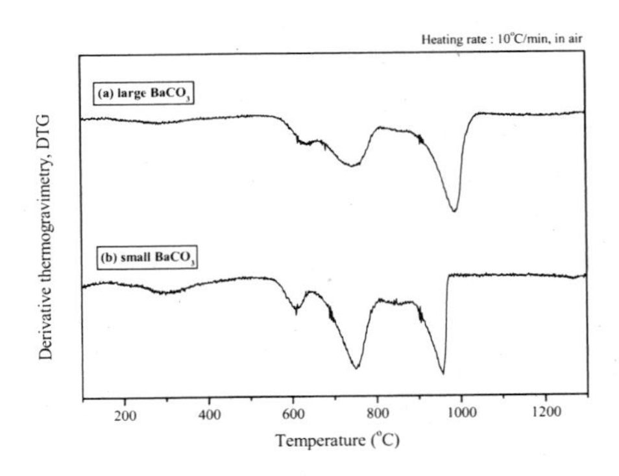

Figure 5. DTG curves of the 1.0 mg/ml PEI added mixture of TiO_2 and $BaCO_3$ with heating rate of 10°C/min, (a) using large $BaCO_3$ (BN1), and (b) using small $BaCO_3$ (BN3).

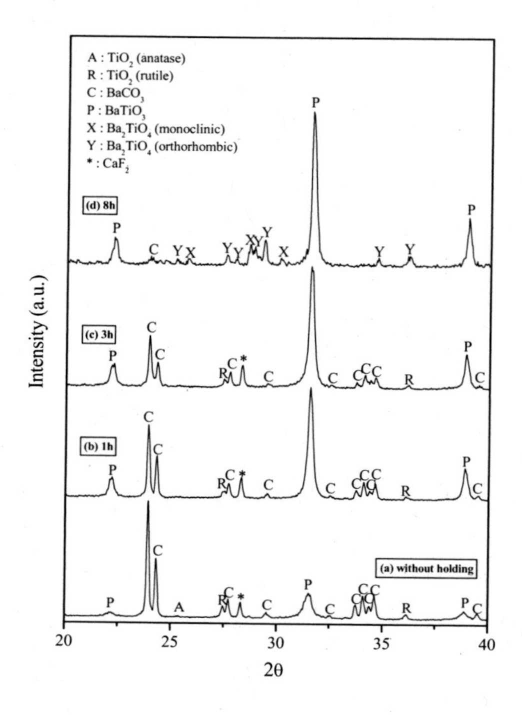

Figure 6. XRD patterns of the mixture of TiO_2 and large $BaCO_3$ (BN1) with 1.0 mg/ml PEI calcined at 800°C, (a) without holding, (b) 1h, (c) 3h, and (d) 8h.

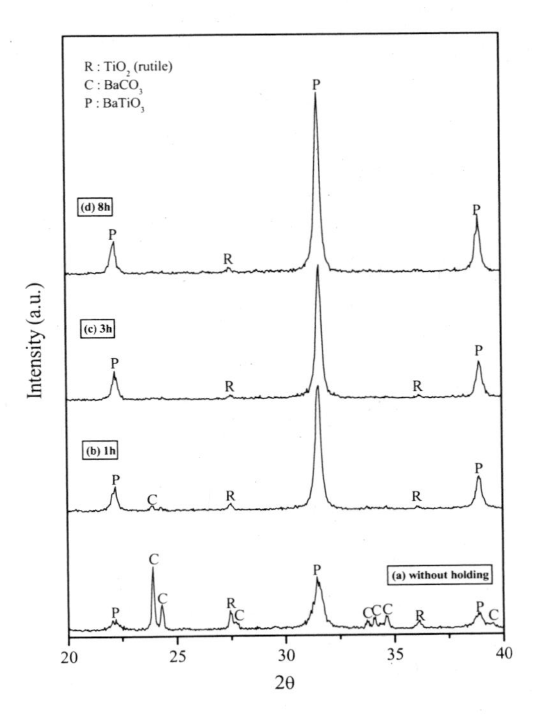

Figure 7. XRD patterns of the mixture of TiO_2 and small $BaCO_3$ (BN3) with 1.0 mg/ml PEI calcined at 800°C, (a) without holding, (b) 1h, (c) 3h, and (d) 8h.

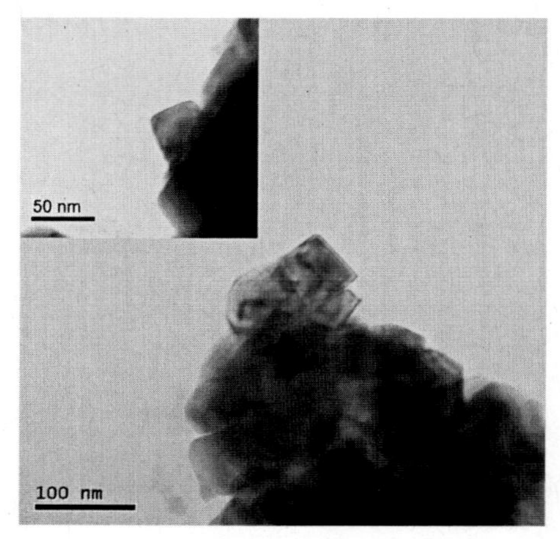

Figure 8. TEM photographs of the mixture of TiO_2 and small $BaCO_3$ (BN3) with 1.0 mg/ml PEI calcined at 800°C for 4h.

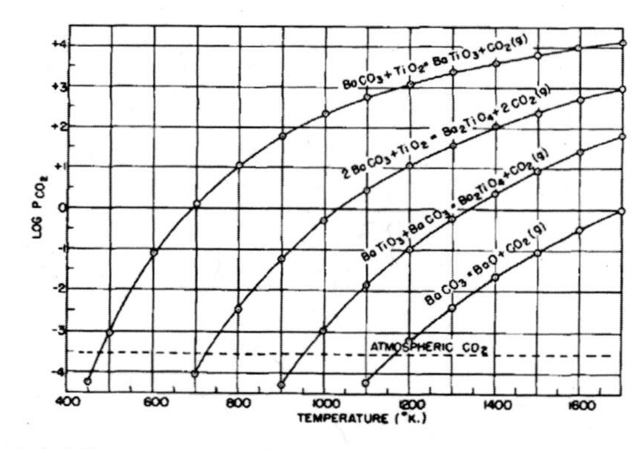

Figure 9. Calculated CO_2 pressure as a function of temperature for the indicated reactions.[26]

MORPHOLOGICAL STABILITY OF GOLD NANOPARTICLES ON TITANIA NANOPARTICLES

M. Nahar,[a] and D. Kovar*[a,b]

[a]Materials Science and Engineering Program
[b]Department of Mechanical Engineering
University of Texas at Austin
Austin, TX 78731

ABSTRACT

Au and TiO_2 NPs were made using the Laser Ablation of Microparticles Aerosol (LAMA) process, which produced bare (i.e. free of surfactant) Au NPs that were 1 – 10 nm in diameter on the surface of amorphous TiO_2 NPs that were ~50 nm in diameter. The stability of Au nanoparticles (NPs) deposited onto TiO_2 nanoparticles was studied using *in-situ* transmission electron microscopy (TEM). Upon exposure to the electron beam, the larger Au NPs were observed to be mobile on the surface of the TiO_2 NPs. However, the smallest Au NPs appeared to be pinned to the TiO_2 surface and resisted sintering. The morphological stability of the small, isolated Au NPs under electron beam exposure suggests that these particles may also be stable at elevated temperatures and therefore would be suitable as high surface area catalysts for oxidation of CO.

INTRODUCTION

Nanoparticles (NPs) have applications spanning various disciplines including catalysis, drug delivery, optoelectronics, electronic devices and others. In addition to simple one component NPs, there is considerable interest in multi-component NPs that contain more than one material constituent [1] because they offer the potential to tailor the chemistry and electronic structure of materials. This is particularly important in the design of catalysts [2]. For example, Goodman *et al.*[3] have shown significantly enhanced catalytic activity for CO oxidation in 1 – 2 nm gold NPs when they are in contact with a flat titania surface. This increased catalytic activity of Au NPs on a TiO_2 surface occurs for a very narrow size range of Au NPs. In has been previously shown that isolated Au NPs tend to sinter at elevated temperatures [4], which would be expected to cause a severe reduction in catalytic activity due to the increase in the sizes of the Au NPs [5]. In this paper we study the size-stability behavior exhibited by Au NPs supported on TiO_2 NPs produced by the LAMA process.

EXPERIMENTAL PROCEDURE

The LAMA process for producing single component NPs has been previously reported [6]. For this study, an additional ablation step, shown schematically in Figure 1, was utilized to produce Au NPs on TiO_2 NPs [7]. A particle feeder containing a vibrating membrane attached to a solenoid driven at 10 Hz was used to aerosolize the feedstock particles in a He carrier gas. A coaxial flow of He gas was utilized to ensure a laminar flow of the aerosol through the ablation points. The flow gas carried the aerosolized TiO_2 micro-particles to Ablation point 1, where they were irradiated by a KrF excimer laser (λ= 248 nm, rep rate = 200 Hz) focused midway between Ablation points 1 and 2. Because the laser fluence was above the breakdown threshold of the microparticles, plasma breakdown and the generation of a shockwave resulted. This shockwave compressed and heated the TiO_2 powder particles above the critical point. Condensation of TiO_2 NPs occurred in the rarefaction behind the shockwave, as expansion and cooling occurred. The TiO_2 NP aerosol was then carried to Feeder 2 and mixed with micron-sized feedstock Au particles. The gas flow carried the mixed TiO_2 NP/Au microparticle aerosol to Ablation spot 2. Because the absorption depth for the laser was larger than the TiO_2 NPs, the TiO_2 NPs were heated but did not reach the threshold for breakdown. As a result, the TiO_2 NPs were reduced slightly in size by evaporation but otherwise remained intact. In contrast, a shockwave was generated in the larger Au microparticles that resulted in the formation of Au nanoparticles that condensed on the surfaces of the TiO_2 NPs. The Au/TiO_2 particles were collected onto lacey carbon TEM grids. For comparison, free Au NPs were also produced and deposited directly onto a lacey carbon grid using the same setup by switching off aerosol feeder 1.

A JEOL 2010F TEM equipped with a room temperature stage was operated at 200 kV, to both image the NPs and to examine the influence of the electron beam on the size-stability of the Au NPs. The electron beam intensity on the NPs was 2.8×10^9 – 2.9×10^9 e/cm$^2\cdot$s under conventional high vacuum conditions. Dynamic behavior of the NPs was observed *in situ* at high resolutions.

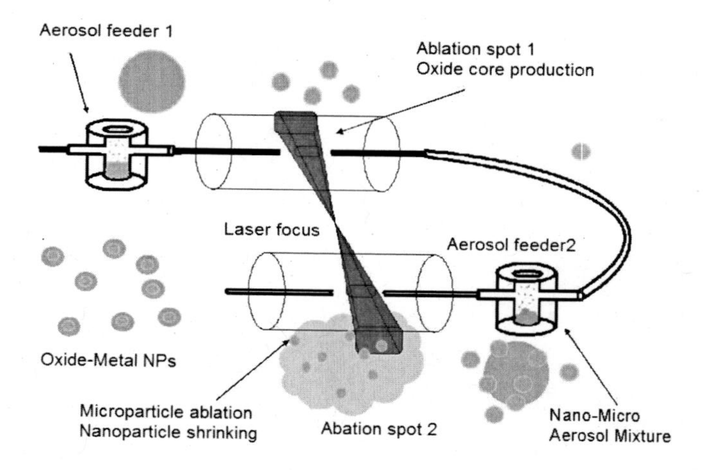

Figure 1: Schematic of the core-shell fabrication technique.

RESULTS AND DISCUSSIONS

Free Au NPs lying on lacey carbon TEM grids were irradiated at an electron beam intensity $2.8\times10^9 - 2.9\times10^9$ e/cm^2·s. Figures 2 and 3 shows the effects of electron bombardment on the size stability of Au NPs of two different size ranges. As expected from conventional sintering theory, the larger gold NPs sintered at a slower rate [8]. Neck formation and sintering were apparent for the larger particles over a time span of 180 s, whereas the smaller particles completely sintered into a single, polycrystalline particle within a span of only 40 s.

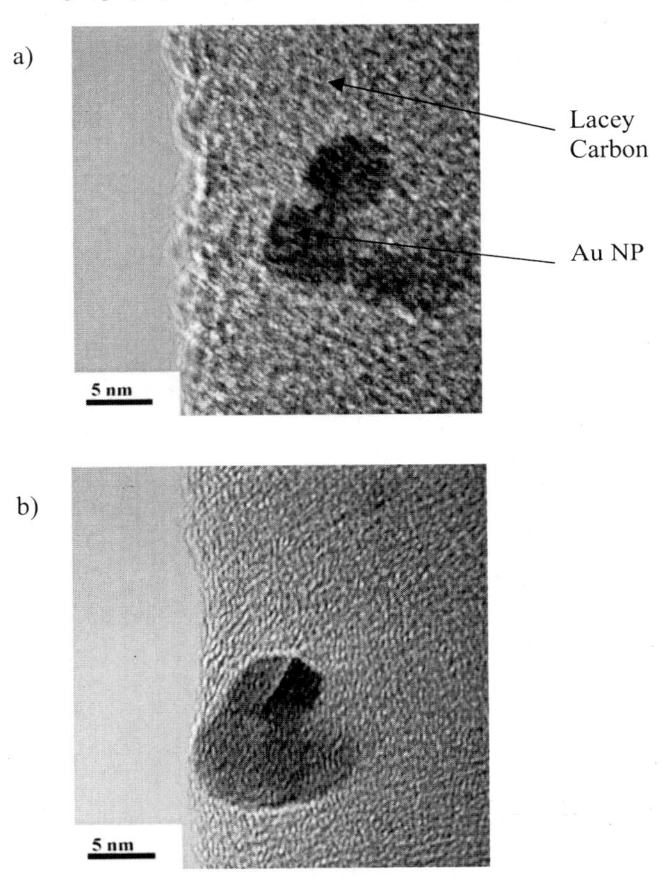

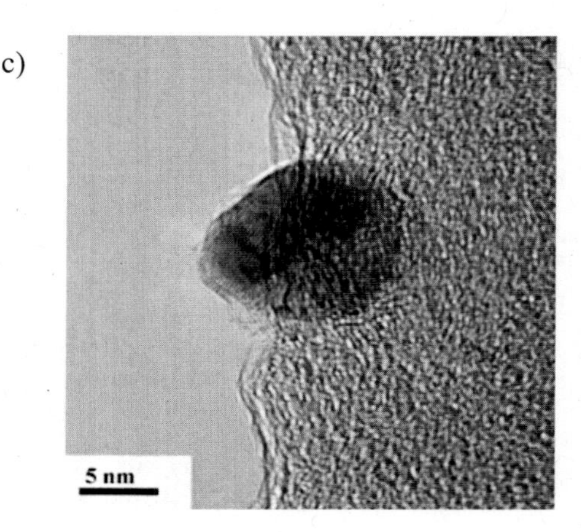

Figure 2: TEM images of larger Au NPs exposed to an electron beam intensity of 2.8×10^9 – 2.9×10^9 e/cm^2·s, after (a) 0 (b) 90 s (c) 180 s.

Figure 4 shows a TEM micrograph of Au/TiO_2 NPs produced by the LAMA process. To study the differences in size stability we deliberately chose a large TiO_2 NP that contained a broad Au NP size distribution. The TEM micrographs in Figure 5 show the time-dependent sintering/ migration behavior of the Au NPs on a titania NP, when irradiated with an electron beam intensity of 2.8×10^9 – 2.9×10^9 e/cm^2·s. The effect of electron bombardment causes migration/sintering behavior that it is not predicted from sintering theory. For example, NP A migrates away from the smaller NP C and moves towards NP B, eventually forming a neck. NP D appears to migrate freely on the surface of titania. NPs C, E, F, G move relative to each other but do not sinter. Certain NPs such as B-D, F-G exhibit desintering. These observations, though not conclusive, give the following atypical results – (1) A few Au NPs appear to be free to migrate on the surface, while others are pinned to the titania surface (2) 1 – 2 nm NPs of Au lying freely on lacey carbon, sinter readily. However, many small Au NPs lying on the TiO_2 NP do not sinter even after long electron beam exposures but are mobile on the TiO_2 NP surface (3) A few NPs show an unusual desintering type of behavior.

One possible hypothesis that could explain the differences in behavior between these Au particles are differences in the way they attach themselves to the titania surface, i.e. whether some of the Au NPs are simply physically attached to the titania NP or whether they are instead chemically bonded to the surface. An alternative hypothesis is that surface energy anisotropy effects are dominant so that the mobility of the NPs depends on the relative orientations of the Au NP on the TiO_2 surface.

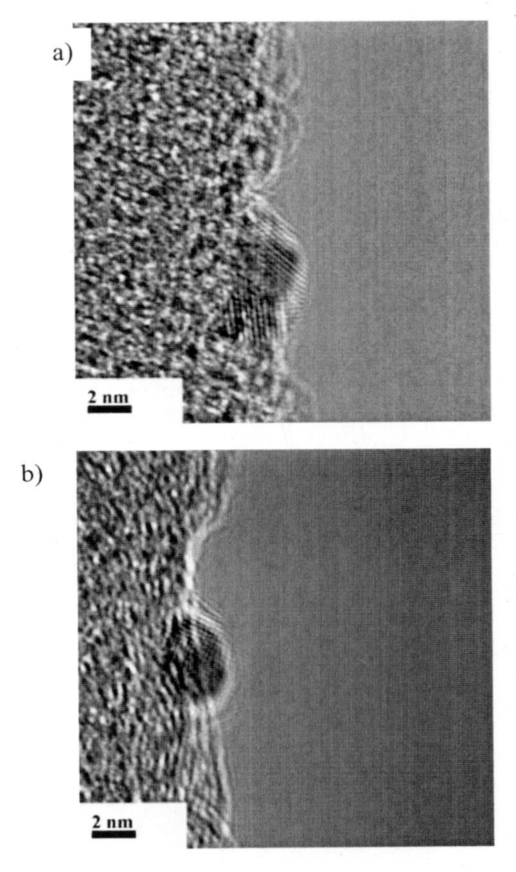

Figure 3: TEM images of smaller Au NPs exposed to an electron beam intensity of 2.8×10^{9} – 2.9×10^{9} e/cm^{2}·s, after (a) 0 s (b) 40 s.

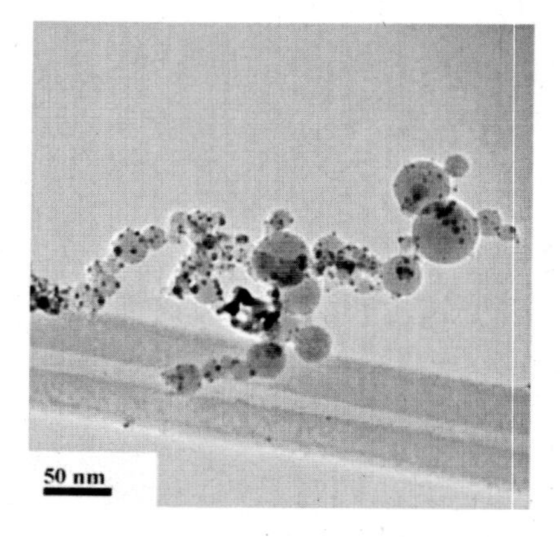

Figure 4: TEM image of Au-TiO_2 NP on NP system produced by the LAMA process.

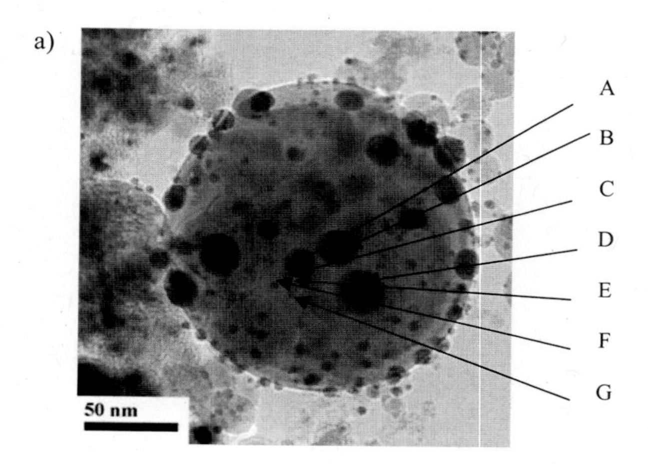

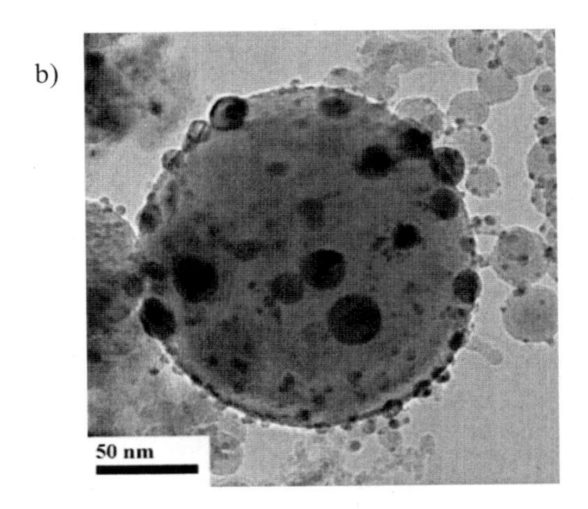

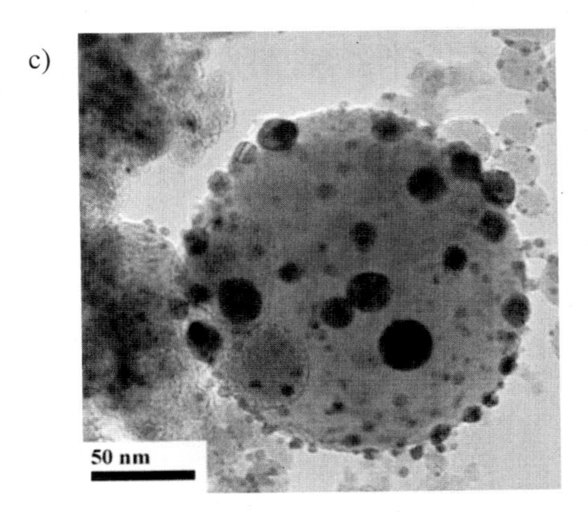

Figure 5: TEM images of Au NPs on TiO_2 NPs exposed to an electron beam intensity of $2.8\times10^9 - 2.9\times10^9$ e/cm^2·s, after (a) 0 s (b) 300 s (c) 450 s.

CONCLUSIONS

We have demonstrated a method to produce bare, metallic NPs on the surface of oxide NPs. Electron bombardment on Au/TiO_2 NPs at room temperature revealed anomalous behavior of the Au NPs when compared to Au NPs that were not on the surface of TiO_2 support NPs. Whereas the unsupported Au NPs sintered freely at room temperature under the electron beam, the TiO_2-supported Au NPs exhibited a range of behaviors that appeared to be correlated to their size. Some of the larger NPs readily sintered while others showed an unusual desintering behavior. The larger Au NPs readily migrated on the surface of TiO_2, whereas the smaller NPs appeared to be pinned to the TiO_2 surface. These observations suggest that the migration of Au NPs on TiO_2 could be dependent on the mode of attachment of Au on TiO_2 or on surface anisotropy effects. Although in these studies the electron beam was used as the means for providing energy to the NPs to study sintering at room temperature, NPs produced by LAMA may also exhibit enhanced morphological stability at high temperatures that would make them good candidates for catalysis. These preliminary results suggest that more detailed and systematic studies are needed to clarify the anomalous sintering behavior for metallic NPs that reside on the surface of larger oxide NPs.

ACKNOWLEDGEMENTS

This work was supported by the National Science Foundation under CBET 0708779. This work was also supported by the Laboratory Directed Research and Development (LDRD) program at Sandia National Laboratories. Sandia is a multiprogram laboratory operated by Sandia Corporation, a Lockheed Martin Company, for the United States Department of Energy's National Nuclear Security Administration under Contract DE-AC04-94AL85000. The authors would like to acknowledge helpful discussions with Mr. Kris Gleason, Dr. Ignacio Gallardo, Dr. John W. Keto, Dr. Michael Becker, and Dr. J.P Zhou.

REFERENCES

[1] E. Arici, D. Meissner, F. Schaafler, and N.S. Sacriciftci, Core-shell Nanomaterials in Photovoltaics, *Int. J. Photoenergy*, Vol 05 (No. 4), 2003, pp 199-208.
[2] C.J. Zhong and M.M. Maye, Core-Shell Assembled Nanoparticles as Catalysts, *Adv. Mater.*, Vol 13, 2001, pp 1507-1511.
[3] M.S. Chen and D.W. Goodman, The Structure of Catalytically Active Gold on Titania, *Science* Vol 306 (No. 5964), 2004, pp 252-255.
[4] Y. Chen, et al., Sintering of Passivated Gold Nanoparticles under the Electron Beam *Langmuir*, Vol 22 (No.6), 2006, pp 2851-2855.
[5] M. Haruta, Size and Support-Dependency in the Catalysis of Gold, *Catal. Today*, Vol 36, (No. 1), 1997, pp 153-166.
[6] J. Lee, M.F. Becker. J.W. Keto, Dynamics of Laser Ablation of Microparticles prior to Nanoparticle Generation, *J. Appl. Phy.*, Vol 89 (No. 12), 2001, pp 8146-8152.
[7] I. Gallardo, K. Hoffman, and J. W. Keto, Applied Physics A: Mat. Sci. and Processing Vol 94, 2009, pp 65-72.
[8] K. Komeya and H. Inoue, Sintering of Aluminum Nitride: Particle Size Dependence of Sintering Kinetics, *J. Mat. Sci.*, Vol 4 (No. 12), 1969, pp 1045-1050.

SYNTHESIS OF NANOSTRUCTURED MESOPOROUS ORDERED SILICA SUPPORTED Fe_2O_3 NANOPARTICLES FOR WATER PURIFICATION

Sawsan A. Mahmoud[1]* and Heba M. Gobara[2]*

Egyptian Petroleum Research Institute (EPRI), (The Process Design and Development Division)[1]and (Refining Division, Catalysis Department)[2], 1 Ahmed El-Zomer St., Nasr City, Cairo, Egypt, P.O.: 11727[1]
*Corresponding Authors Email: sawsanhassan2003@ yahoo.com[1], hebagobara @ yahoo.com

ABSTRACT
Nano-particles Fe_2O_3 supported over SBA-15 was synthesized using the post synthesis method in the ratios 0.6, 0.9, 1.5 and 2 wt % under the effect of visible light. Several characterization techniques of the prepared samples were carried out such as thermal analysis (differential scanning coulometery (DSC) and thermal gravimetric analysis (TGA)), N_2-adsorption-desorption, x-ray diffraction (XRD), fourier transform infrared (FTIR), scanning electron microscopy (SEM) connected with energy dispersive spectroscopy (EDS) and. x-ray photoelectron spectroscopy (XPS). The results showed that all the prepared samples are thermally stable up to 800°C. N_2-adsorption-desorption showed that all the isotherms are of type IV which is characteristic of mesoporous materials with H1 hystersis loop, which indicate the presence of regular cylindrical pores. XRD analysis showed the presence of three well-resolved peaks at (100), (110) and (200) characteristic for the ordered mesoporous structure. Upon loading of iron, the ordered mesoporous structured of SBA-15 was maintained. No diffraction peaks corresponding to crystalline iron oxides were detected. XRD data also showed that no diffraction peaks corresponding to crystalline iron oxide were detected. This depicted that Fe_2O_3 was incorporated into the silica framework or the Fe_2O_3 was highly dispersed on the surface of SBA-15. The XPS results ensured the incorporation of Fe_2O_3 into the silica framework in all the samples studied except one of 0.9 wt %, where the Fe_2O_3 phase was found on the surface in a highly dispersed state. The prepared samples were tested in the photocatalytic degradation of phenolsulfonphthalein (PSP) under visible irradiation. The present work emphasized that such preparation method and the incorporation of the nanoparticles Fe_2O_3 into the silica framework with preserving the ordered structure of mesoporous SBA-15 have a great effect in the photodegradation of phenolsulfonphthalein.

1. INTRODUCTION

Nanosized materials have received significant attention in application fields as diverse as electronics, nonlinear optics, medicine, chemistry and engineering, e.g., for preparation of high density magnetic recordings, magnetic fluids, solar cells, catalysts, sensors, membranes, etc. [1–4]. The unique properties of nanoparticles originate from their high surface to volume ratio, i.e. by the dominant behavior of the surface atoms, which gives the materials special properties, such as super paramagnetism, a blue shift of the optical absorption edge, and remarkable catalytic and reductive properties [5–12]. Very often, the nanoparticles are deposited onto porous carriers. However, the level of nanoparticle dispersion on the carrier is influenced by a number of parameters, such as the method of nanoparticles deposition, the loading degree, the pore structure, pore diameter, and pore volume of the support, and the interaction between the support and the nanoparticles [13–19]. Among the porous materials, the mesoporous silica materials involving uniform pore sizes and high surface areas have been widely applied as a host for loaded catalysts [20], polymers [21], metals [22] and semiconductors [23]. Compared with MCM-41, the SBA-15 material possesses a high surface area and uniform tubular channels with tunable pore diameters in the range of 5– 30 nm, which are significantly larger than those of MCM-41. Especially due to its thicker walls, SBA-15 provides thermal and hydrothermal stabilities that exceed those for MCM-41 with thinner walls. Additionally, cheaper surfactant was used during its preparation. Obviously, SBA-15 is a more ideal catalytic support than MCM-41.

Li and co-workers revealed that mesostructure was still maintained when iron oxide nanoparticles were supported onto SBA-15 at mild acidic conditions [24]. Fröda and co-workers reported SBA-15-supported Fe with different precursors in dry ethanol via post-synthesis procedure and checked also the direct hydrothermal methods [25]. They found that different strategies of synthesis lead to significant changes in the bonding and the environment of iron species within the silica materials. Nevertheless, some reporters showed that the nanoparticles supported on SBA-15, by impregnation or grafting methods are not suitable to prepare catalysts with highly isolated catalytic active sites.

Recent studies have demonstrated that heterogeneous photocatalysis using TiO_2 appears as the most emerging destructive technology [26]. However, the large band gap energy (3.2 eV) for anatase TiO_2 (excitation wave length <387.5 nm) limits its practical application under the condition of natural solar light [27]. To develop more solar light- efficient catalysts, it seems of urgency to develop photocatalytic systems which are able to operate effectively under visible light irradiation. Iron-containing SBA-15 catalyst consisting of crystalline hematite supported onto a meso-structured silica matrix has been shown as a promising catalyst for the treatment of phenolic solutions through photo-Fenton processes [28], CdSeZn-SBA-15 was used as photo catalyst for the generation of H_2 from 2-propanol aqueous solution under UV light irradiation [29]. Besides, TiO_2-loaded Cr-modified SBA-15 was prepared for 4-chlorophenol photo-degradation under visible light [30]. Moreover, 2, 9-dichloroquinacridone sensitized Ti-SBA-15 (DCQ-Ti-SBA-15) was employed to decompose indigo carmine displaying high photocatalytic efficiency under UV light irradiation [31]. The phenolsulfonphthalein dye (PSP) was chosen in this study as a pollutant in the photocatalytic degradation test using visible light. Such dye is used frequently for estimation of overall blood flow through the kidney.

The main purpose of our work is to synthesize mesoporous silica containing highly dispersed nano-scaled Fe_2O_3 particles by sol-gel method. Investigation of the prepared catalysts by various characteristics adopting several techniques, viz., XRD, N_2 physi-sorption, FT-IR SEM, XPS and EDEX. The prepared catalyst samples were applied in the photocatalytic degradation of $PhSO_3Ph$ (PSP) using visible light. The catalytic activities of the investigated catalyst samples were attempted to be discussed in correlation with their various characterization parameters.

2. EXPERIMENTAL

2.1. Synthesis of SBA-15 and Fe-SBA-15 catalysts

Mesoporous SBA-15 silica was prepared according to the literature using tri-block polymer pluronic P123 ($EO_{20}PO_{70}EO_{20}$; Aldrich) as surfactant in acidic conditions [32]. In a typical synthesis, 4 g Pluronic P123, ($EO_{20}PO_{70}EO_{20}$; Aldrich) triblock copolymer was dissolved in 30 g distilled water. A 2 M HCl (120 g) solution was slowly added to the solution under stirring at 40°C, and 8.5 g of tetraethylorthosilicate (TEOS; Aldrich) was then added. The resulting solution was stirred for 20 h at 44 °C, followed by aging at 80°C for 24 h under static conditions. The solid product was washed with distilled water, recovered by filtration and dried at room temperature. The template was removed from as-made mesoporous material by calcination at 550°C for 6 h (heating ramp = 1°C min^{-1}). This calcined support was denoted as SBA-15.

The iron oxide was deposited onto mesoporous SBA-15 by classical wet impregnation using aqueous solution of iron acetyl acetonate dissolved in 50 ml of absolute ethanol and stirred for 30 min. One gram of SBA-15 was added to the solution while stirring for 30 min and then irradiated by visible light using halogen lamp for 20 min. The composite was separated from the solution by centrifugation (≥ 5000 rpm) and dried under vacuum (40°C, 3h). Finally the samples were calcined at 450 °C for 3h. The obtained solid products, designated as I, II, III and IV, contained 0.6, 0.9, 1.5 and 2 wt % Fe_2O_3, respectively, as evidenced from EDX measurements.

2.2. Characterization

The solid materials were characterized by powder X-ray diffraction recorded on a Brucker D8 advance X-ray diffractogram with Cu Kα radiation (λ = 1.5418 Å). N_2 adsorption-desorption isotherms at -196 °C were obtained with a NOVA 3200 apparatus, USA. The samples were previously outgassed under vacuum at 200 °C for 4 h. Surface areas (S_{BET}) were calculated from multi-point at relative pressure (P/Po) ranging from 0.05 to 0.30. Particle size distribution was calculated from Barrett, Joyner and Halenda (BJH) method using adsorption branch of the isotherms. IR experiments were performed using AT1 Mattson model Genesis Series (USA) infra red spectrophotometer. For all samples, the KBr technique was carried out approximately in a quantitative manner, as the weight of the sample and that of KBr were always kept constant.

DSC-TGA analyses were carried out for all supported metal oxide samples using simultaneous DSC-TGA SDTQ 600, USA under N_2 atmosphere, with a heating rate of 10°C min^{-1}. The morphology of the samples was studied by the aid of scanning electron microscope, JEOL – JSM-5410, coupled with energy dispersive X-ray spectroscopic (EDX-Oxford) analyzer working at 20 KeV to elucidate the distribution mode of Fe and Si elements. All the samples were analyzed under moderate vacuum after gold coating. XPS was carried out using Omicron Surface Instruments, USA. Al probe with 150 Watt power was used. The charge correction was done by hydrocarbon peaks at 284.6 eV.

The catalytic activity of the various samples was tested in photodegradation of PSP using visible light according to the following procedure: 500 cc of an aqueous solution containing 50 ppm of highly pure (PSP) was subjected to visible irradiation using a 125 Watt halogen lamp. All the experiments were conducted in a batch reactor. The halogen lamp was placed in a cooling silica jacket and placed in a jar containing the polluted water. The catalyst suspension (0.1 g/L) was stirred in the solution using a magnetic stirrer at a controlled reaction temperature of 20°C during the experimental period. At different irradiation time intervals, samples of the irradiated water were withdrawn for analysis using Perkien Elmr series 200 HPLC photodiode array detector and extract the chromatogram at 530 nm wave length.

3. RESULTS AND DISCUSSION

3.1. X-ray Analysis

Figure 1 (a) shows the XRD patterns of pure SBA-15 and Fe-SBA-15 samples with different Fe concentrations. The XRD patterns for all samples in low-angle region show three well-resolved peaks at (100), (110) and (200), which are characteristic of mesoporous material with 2D-hexagonal structure. They resemble that one of the parent mesoporous silica SBA-15, indicating that the materials prepared have highly ordered mesoporous structures [32]. Moreover, in the wide-angle region the wide line of amorphous silica was observed, no diffraction peaks corresponding to crystalline iron oxides could be observed (Fig. 1b). Possibly, iron species were introduced into the framework of ordered mesoporous silica or the particle size of iron oxides in the sample was too small to be detected by x-ray diffraction. The absence of these prominent reflections in case of the nanostructured Fe_2O_3 indicated that no crystalline bulk materials have been formed outside the pore system. The result also showed that Fe_2O_3 nanoparticles were dispersed uniformly in the frame of SBA-15 [33].

Since the radius of the ionic Fe^{3+} is larger than that of Si^{4+} ($r^{3+}{}_{Fe}$ = 0.63 Å and $r^{3+}{}_{Si}$ = 0.40 Å), the increase in the αo values (Table 1) may indicate that Fe^{3+} ions are incorporated into the framework of the SBA-15 materials [34]. Generally, it can be expected that the unit cell parameter should be enlarged after the incorporation of metal cations with ionic radius larger than that of Si^{4+}.

3.2. N_2 Physisorption

N_2 adsorption–desorption isotherms and BJH pore size distributions for all samples are illustrated in Fig. 2. The derived surface data are collected in Table 1. The deposition of Fe species on SBA-15 support did not modify greatly the isotherm shape of SBA-15 support. The obtained isotherms for SBA-15 and Fe-SBA-15 samples are of type IV according to the IUPAC classification [35], being typical for mesoporous materials with 2D-hexagonal structure that have

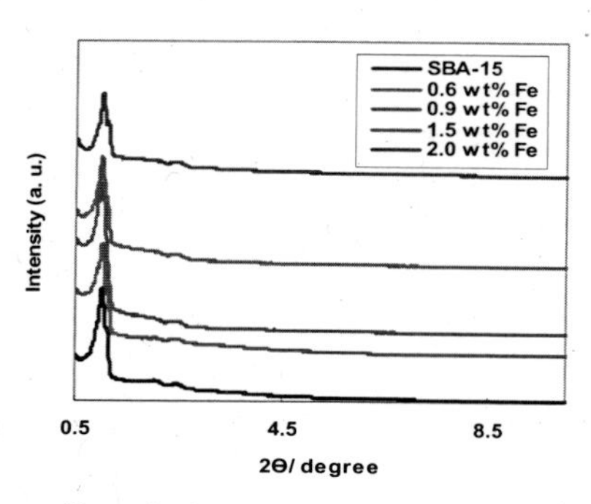

Figure 1a: low angle XRD of Fe-SBA-15 catalysts.

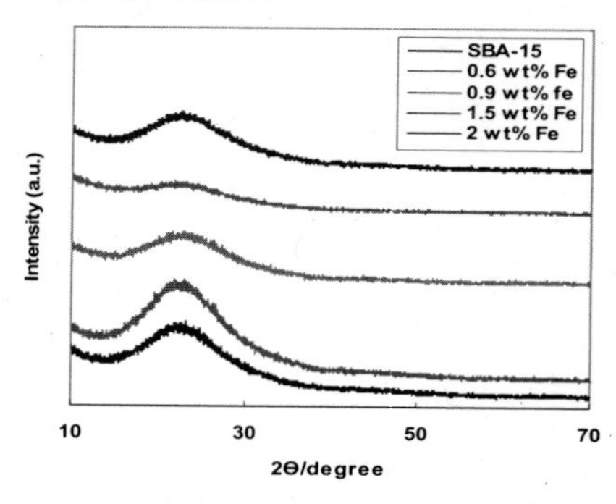

Figure 1b: high Angle XRD of SBA-15 support and Fe-SBA-15 catalysts.

large pore sizes with narrowed pore size distributions [32]. The observed H1 type hysteresis loops are typical of mesoporous materials with uniform tubular pores [36, 37]. Such H1 type hysteresis loops indicate also the presence of open channels. The well-defined step, occurred at high relative pressures (0.4–0.8), correspond to capillary condensation of N_2, related to the uniformity of the pores.

Table 1: Physicochemical properties of SBA-15 and Fe-SBA-15 materials with different Fe contents.

Catalyst Samples	αo^{a} (nm)	S_{BET} (m^2/g)	Vt (cm^3/g)	W_{BJH} (nm)	$T_W{}^{b}$ (nm)	Rate of photo catalytic degradation, min^{-1}
SBA-15 support	9.96	707	0.741	4.036	5.93	
Sample I	10.10	742	0.847	4.030	6.07	0.043
Sample II	10.17	537	0.658	4.044	6.12	
Sample III	10.09	614	0.666	4.038	6.05	0.043
Sample IV	9.99	621	0.696	5.58	4.41	0 .084

αo : The length of the hexagonal unit cell = $2d_{100}/(3)^{1/2}$.

T_W: The wall thickness = αo - W_{BJH}.

For the sample II, the surface data are close to those of pure support. A part of the oxide seems to be incorporated within the channels of SBA-15 as accompanied with some increase in the αo values. Also, a marked decrease in S_{BET} and the total pore volume could be observed due to the incoroporation of iron oxide particles. This is in agreement with the observations reported by Vradman et al. for TiO_2 [38]. The main fraction seems to exist most likely at the outer surface of the support. By increasing Fe_2O_3 lading to 0.9 wt %, a marked change can be observed where the fraction incorporated increased, as accompanied with an increase in αo value and the wall thickness (Tw). Some thickening of the walls and pore blockage of some of the channels can be reported for this sample. The outside surface of the channels is leveled up. This indicates that finely dispersed metal oxide particles locate predominantly on the surface and inside the channels for this catalyst sample [39]. The S_{BET} and the total pore volume values decrease gradually up to 1.5 Fe_2O_3 wt%. The extent of incorporated fraction of Fe_2O_3 remains almost unchanged. For the sample of 2.0 wt% loading, only little increase in S_{BET} and pore diameter is observed. Figure 2B reveals the appearance of a new pore system with some wider pore diameter. A more fraction of the oxide seems to be located free on the outer surface as result of the diffusion from the pore system to the outer surface of SBA-15, as revealed by the open channels.

S_{BET} and the total pore volume values gradually decrease up to 1.5 wt%. The extent of incorporated fraction of Fe_2O_3 remains almost unchanged. For the sample of 2.0 wt% loading, a little bit increase in S_{BET} and pore diameter is observed. Figure 2B reveals the appearance of a new pore system with little bit wider pore diameter. Amore fraction of the oxide seems to be located free on the outer surface as result of the diffusion from the pore system to the outer surface of SBA-15 since we have open channels.

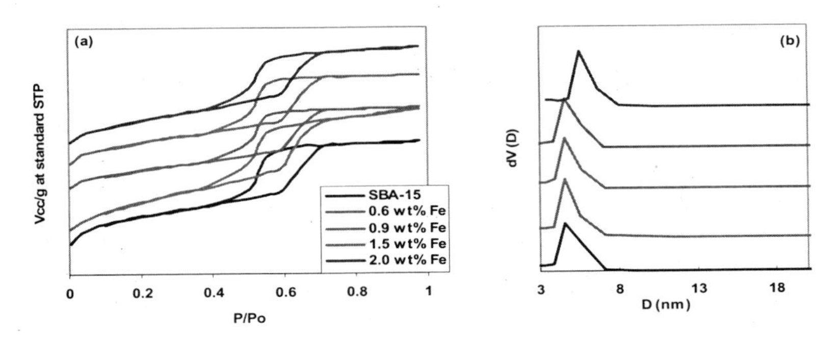

Figure 2: N_2 adsorption-desorption isotherms (a) and pore size distribustion (b) of SBA-15 and Fe-SBA-15 samples.

3.3. FT-IR Analysis

Figure 3 shows FT-IR spectra in the scan range from 400 to 4000 cm^{-1} for the parent SBA-15 and various supported iron oxide samples. Peaks at 1065, 797 and 446 cm^{-1} for the parent SBA-15 are assigned to characteristic vibrations of Si-O-Si bridges cross- linking the silicate network [39, 40]. The shift of these peaks to higher frequencies suggests an increase in rigidity of Si-O-Si network as metal content increases [41, 42]. The band at 1065 is due to asymmetric stretching of Si-O and the band at 800 cm^{-1} is assigned to symmetric vibration of Si-O [41, 42].

In both amorphous and crystalline silicates, the ≡ Si-OH stretching of the surface silanol groups gives rise to a band at 960 cm^{-1} [43-46]. This band at 960 cm^{-1} in most of the ordered mesoporous materials indicates the presence of perturbing or defect groups. It could be attributed to asymmetric stretching of Si-O bond neighboring surface silanol group in our parent SBA-15 (Fig.3). This band appeared at 960 cm^{-1} degenerated to a shoulder at 1065 cm^{-1} band in all Fe-containing SBA-15 catalyst samples, which may indicate that more surface silanol groups are consumed by loading the iron species in the preparation step. The surface silanol groups seem thus to be involved in the interaction with the guest species. Very similar band was also observed in titanium siloxane polymers [47], mixed oxides [48] or

TiO_2-grafted on silica [49], which could be attributed to a modification of SiO_4 units indirectly related to the presence of hetero metals. The IR spectra of the Fe-containing samples show a lack of Fe-O stretch vibration band at 576 cm^{-1} [50, 51]. One may recall the obtained XRD results confirming that iron species are incorporated into the framework of SBA-15 support.

For SBA-15, the spectrum consists of a number of bands in the C-H stretching region of 2800-3000 cm^{-1}, being attributable mainly to CH_3 and CH_2 stretching vibrations. This runs in agreement with the fact that ethylene oxide chains occluded in silica walls should remain unchanged [52]. The broad bands ranging from 3200 to 3600 cm^{-1} present in the spectra of all materials are assigned to O-H stretching vibrations of adsorbed water molecules and surface hydroxyl groups strongly perturbed by hydrogen bonding.

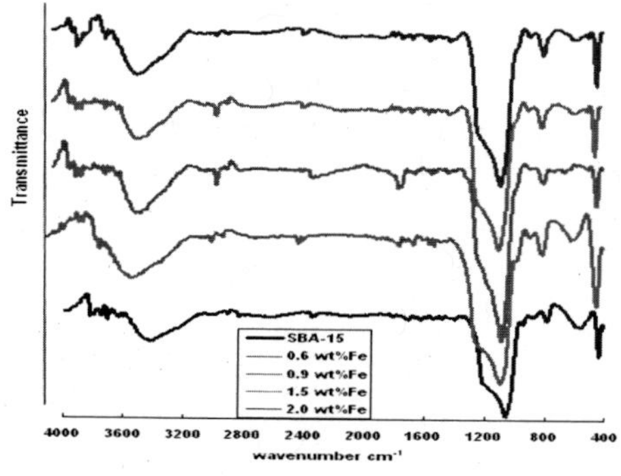

Figure 3 : FT-IR curves of SBA-15 and Fe-SBA-15 catalysts

3.4. Energy Dispersive X-ray Analysis (EDX)

The EDX investigation revealed that Fe_2O_3 content in the various prepared samples are 0.6, 0.9, 1.5 and 2.0 wt%. These ratios coincide with the Fe/Si ratios of 0.002, 0.007, 0.013 and 0.015 used in our impregnation procedure.

3.5. Thermal Analysis (DSC-TGA)

The obtained thermograms for the SBA-15 support and the various Fe-SBA-15 catalysts (Fig. 4) indicate that all the prepared samples are thermally stable up to 800 °C. Only one endothermic peak is observed in all cases at 90 °C as accompanied with weight losses ranging between 4.83 and 6.07 % (cf., TG curves), being related mainly to the removal of physisorbed water. No sign of presence of any peak related to Fe phases or the dehydroxylation of OH groups of the support can be detected. It is evident that the structure of the support is maintained even after the loading of iron by the mentioned sol gel method of preparation. This again may confirm the incorporation of iron species into the SBA-15 support, supporting the XRD and surface data.

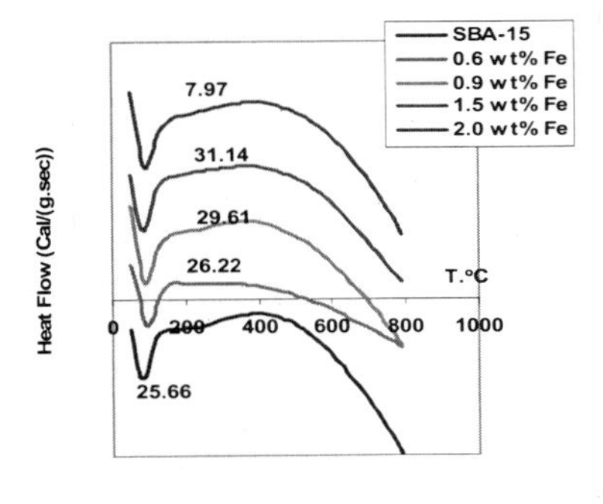

Figure 4a : DSC curves of SBA-15 support and Fe-SBA-15 catalysts.

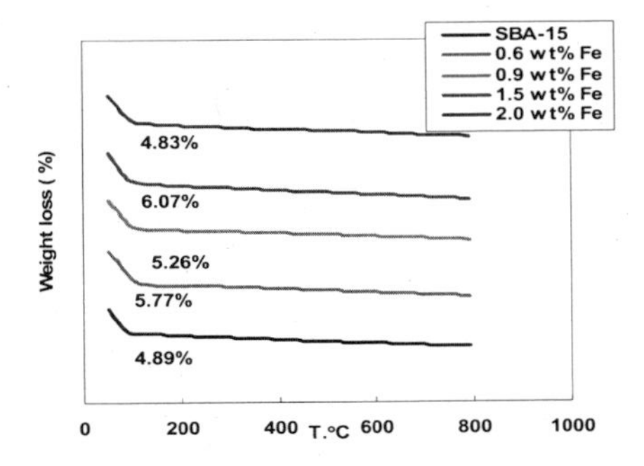

Figure 4b : TGA curves of SBA-15 support and Fe-SBA-15 catalysts.

3.6. Scanning Electron Microscopic Study (SEM)

The morphology of the different samples was investigated through SEM technique. Representative micrograph of SBA-15 support shows the characteristic rod-like morphology [53] (Fig. 5). It is clear that there are no changes in the shape of silica fibers forming the SBA-15 after iron impregnation by the mentioned method of preparation. No particles with a shape different from the packed parallel fibres were detected. This may suggest that the preparation technique applied in this work did not cause degradation of the parent SBA-15 structure. It may also confirm the XRD results that the ordering of the hexagonal array is maintained. The morphology seems also to be consistent with the conclusion of the exclusive incoroporation of iron phase into the framework of the SBA-15 materials, especially in the sample II. Sample II exhibits the highest incorporation extent within the SBA-15 channel, accompanied with a pronounced wall thickness.

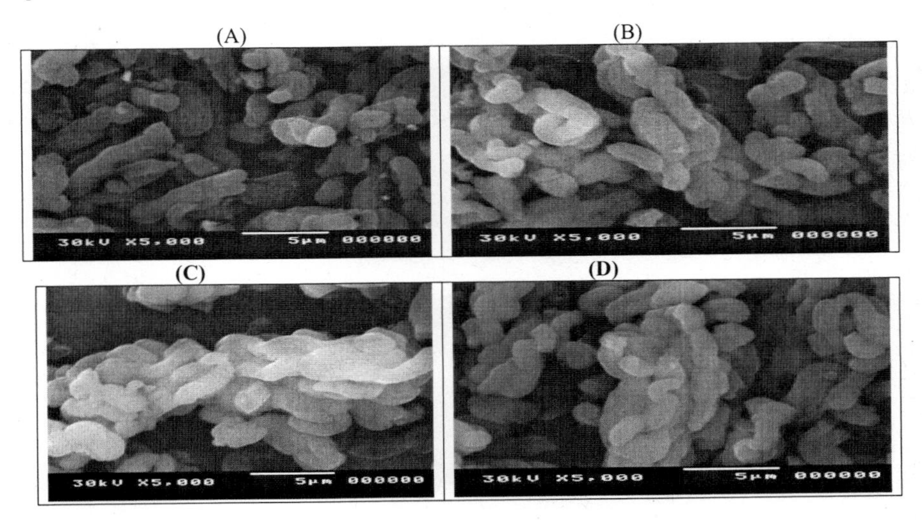

(E)

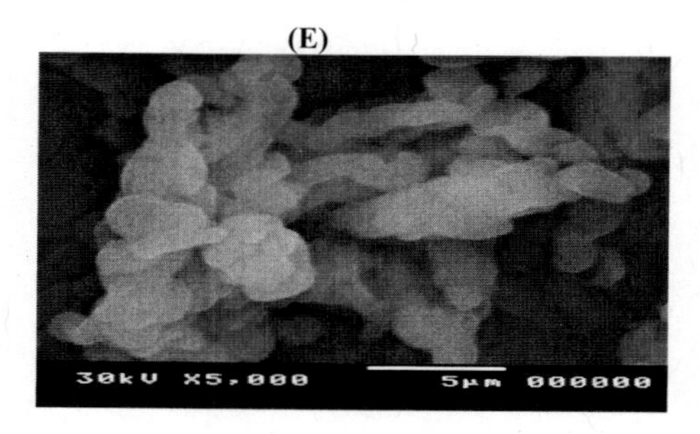

Fig. 5: SEM micrographs of SBA-15 (A) and Fe.SBA-15 catalysts wit different Fe concentration: sample I (B), sampleII (C), sample III(D).and sample IV(E).

3.7. X-ray Photo Electron Spectroscopic Analysis (XPS)

The XPS patterns of the studied samples of different iron loadings, namely I, II, III and IV are illustrated in Fig. 6. For the samples I, III and IV, the obtained patterns showed a characteristic peak at around 706 eV representing most probably iron metal incorporated favorably and stabilized in SBA-15 framework [54]. Another weak peak appeared at 715 eV may represent surface Fe^{3+} cations diffused to the extra framework [55]. The pattern of the sample II can be distinguished from the other ones by the appearance of two peaks of the binding energies at 710 and 723eV, with an energy difference of 13 eV. These two peaks indicate most likely the α- Fe_2O_3 tetrahedral phase, incorporated within the channels of SBA-15 [56], leading thereby to a marked decrease in the specific surface area (Table 1). The exceptional structure of sample II as described should be reflected negatively in the catalytic application as will be shown below. In connection with this observation, it is of interest to recall the oxidation reaction taking place during the calcination step in air. An adsorption –induced surface reaction and a diffusion– induced internal reaction seem to be competing. In the sample II, the formation of iron oxide as entrapped phase in the channel system may be favored by the diffusion-induced internal reaction, as ensured by XPS and surface data. This entrapped phase is not expected to take part in the photocatalytic reaction under investigation.

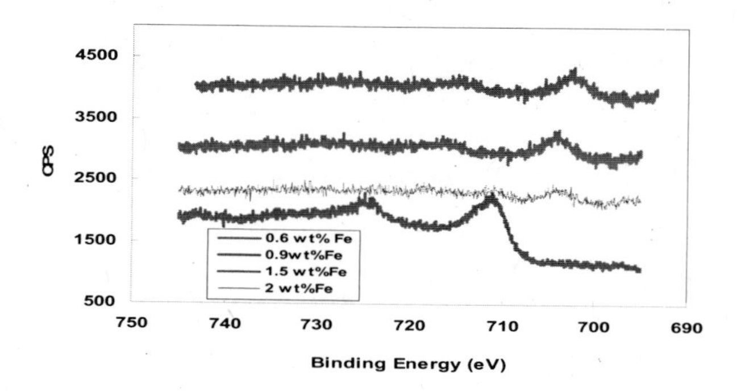

Figure 6: XPS of Fe-SBA-15 catalysts.

3.8. Photocatalytic Activity of the Investigated Samples

The catalytic activity of the various prepared samples was determined in the photo degradation of PSP pollutant under visible irradiation conditions, applying λ at 400- 800 nm. The direct photolysis of PSP indicates a negligible degradation when illuminated with visible light in absence of catalysts. Figure 7a shows the photo degradation of PSP in terms of % degradation as a function of irradiation time in presence of different catalyst samples. The results show that the SBA-15 support exhibits low photoactivity. After 10 min of irradiation, the concentration of the pollutant reaches 45 ppm. beyond which no degradation could be observed up to 70 min, indicating the attainment of equilibrium.

The samples I, III two–stages behavior, firstly, the diffusion into the porous structure of the catalyst where they react at the active sites. Secondly, Products diffuse out of the particle in the opposite direction. The presence of Fe^{3+} and other metallic ions in solution are reduced by free electrons which then accelerate the degradation of organic compounds [57]. The values of the rate of photo catalytic degradation in the range 30-60 min are included in Table 1. The higher activities of these samples may be referred to the metallic Fe incorporated in the SBA-15 framework. On the other hand, the higher activity observed for the samples III (30-60 min) and IV (approaching to 100% degradation) may in addition be linked with the presence of surface Fe^{3+} species diffused to extra framework, as evidenced from XPS and surface data. The sample II, as expected, showed complete inactivity (Fig.5) which confirms the negative effect of Fe^{3+} species (α- Fe_2O_3) located within the SBA-15 channels in tetrahedral settings (cf. XPS and surface data).

However, the diffusion within the SBA-15 mesopores can appear as the rate-limiting step, reducing the catalyst efficiency.

The SBA-15 materials are characterized by regular and uniform mesopore structure. It is a matter of fact that the different catalysts wt % of iron contains mesopores of about the same pore radius (Table 1). However, the the length of the hexagonal unit cell (ao) and the wall thickness (TW) of sample IV is significantly different. Table 1 shows that sample IV has a lower values of ao and Tw than the other samples. The decrease in ao and Tw decrease the diffusion in the interior pores and the diffusion out of the pollutant in the opposite direction of the catalyst and increase the efficiency of the catalyst. Sample II and III have a higher values of ao and Tw so the diffusion into the porous increases and consequently retard the efficiency of the catalyst especially in the first stage (0-30 min). The photon supply poses another limitation to the activity of the internal volumes of the particles .The rate constant includes the concentration of active sites. The active sites are generated by photon absorption. The light intensity decreases exponentially with the penetration depth within the catalyst particle according to the Lambert-Beer law. Again, the internal volume of the largest length hexagonal unit cell receives the lowest photon flux.

It should be noted that a very wide range of optimum iron content can be found in the literature (0.05–20 wt %). Its optimal value might depend on various parameters, such as the synthesis method, conditions, and also on the substrate itself. Zhang et al. [58] found a very interesting correlation between the optimum iron content and the particles size of the photocatalyst. They found that larger nanoparticles require lower iron concentration to reach their maximum photocatalytic activity. Zhang et al. observation is probably not generally true due to the very

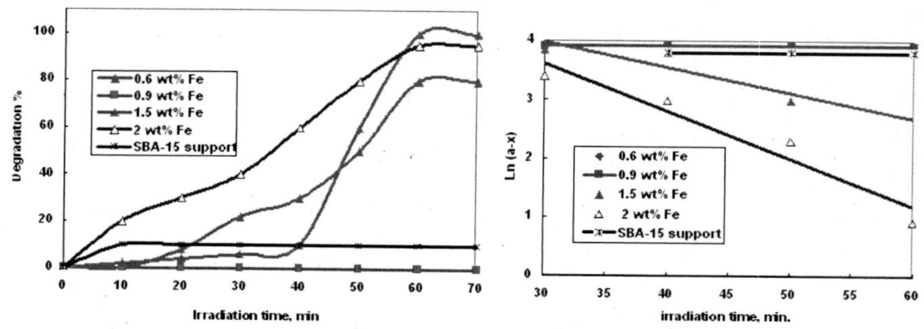

Figure 7a: Photocatalytic degradation of PSP using different wt% of iron species on SBA-15.

Figure7b: First order plot for the photocatalytic Degradation of PSP using different iron species on SBA-15 photocatalyst.

complex nature of the heterogeneous photocatalysis. The dual role of iron in the photocatalytic processes was described by [59-61], in terms of electron (Eqs. 1 and 2) and hole trap (Eqs. 3 and 4) steps:

$Fe^{3+} + e^{-} \rightarrow Fe^{2+}$ (1)

$Fe^{2+} + O_{2ads} \rightarrow Fe^{3+} + O_2^{\cdot -}$ (2)

$Fe^{3+} + {}_{hv}^{+} \rightarrow Fe^{4+}$ (3)

$Fe^{4+} + OH^{-} \rightarrow Fe3^{+} + OH^{\cdot}$ (4)

Since iron (III) ions can act as electron and hole trap and they can release both charge carriers to the solution phase (Eqs. 2 and 4) or directly to the substrate, the recombination rate of the electron–hole pairs is lowered, therefore the photocatalytic activity is enhanced. It should be noted, however, that at higher iron concentration, the iron (III) ions may act as recombination centers as well [61]:

$Fe^{3+} + e- \rightarrow Fe^{2+}$ (5)

$Fe^{2+} + hv \rightarrow Fe^{3+}$ (6)

Since a competition exists among these redox processes, a certain amount of iron (III) ion can work with the best efficiency.

Mechanism of the photocatalytic degradation of PSP under study:

The HPLC chromatogram at 254 nm showed the presence of a photo-intermediate product, which was identified using a standard sample to be phenol. Through analysis of the irradiated liquid product using a buffer and acetonitrile, a peak was detected, which could be identified as sulphonic acid. This may indicate the -OH attack against the carbon benzene rings. Such results were obtained by using the prepared samples SBA-15, I and III but not with the sample IV. The analysis of the HPLC at 254 nm for sample sample IV showed that no aromatic intermediates could be detected by HPLC analysis, indicating the destruction of the benzene ring. The photo- degradation of PSP using the prepared samples may thus obey the following mechanism:

According to Scheme 1, the variation of the catalytic activity of the prepared samples can be interpreted. The retardation of the degradation processes the samples I and III, may be due to the adsorption of phenol on the surface of the catalyst.

The photocatalytic degradation of PSP using the sample IV was faster than using the samples I and III, as evidenced by the absence of aromatic intermediates. The

A simple kinetic treatment of the photodegradation data of PSP using the different catalyst samples under investigation is depicted in Fig. 7b, where the first order kinetics is realized and the calculated rate constants (kcp) are summarized in Table 1.. The rate constant for samples IV, III and II were 0.081, 0.043 and 0.04 3 min^{-1} respectively at (30-60 min).

CONCLUSION

Nanostructured mesoporous ordered silica (SBA-15) containing highly dispersed nano-scaled Fe_2O_3 particles, with different weight percentages, was prepared by a sol-gel method. The prepared samples were investigated by various characterization techniques, viz., N_2 adsorption-desorption, XRD, FT-IR, DSC-TGA, SEM connected with EDX and XPS. All the samples were thermally stable up to 800°C. The XRD results showed the uniformity of the ordered hexagonal structure even after loading with iron species by the mentioned method of preparation. The results for the samples I, III and IV (of 0.6, 1.5 and 2.0 wt% Fe_2O_3, respectively) revealed that the iron was incorporated into the support framework, encouraged most probably by the effect of visible light applied during the loading process. From the XPS results for the samples I, III and IV, the supported iron was found to exist as metallic species in the framework of the SBA-15. For the sample II, iron, especially in the form of α- Fe_2O_3, was incorporated within the support channels, as characterized by the lowest surface area and the largest wall thickness. The photocatalytic activity of the prepared samples was estimated toward the degradation of PSP dye. Both the photodegradation rate and pathway depended on the iron loading and its mode of incorporation into the SBA-15 support. The complete mineralization was evident by using the sample III, while sample II had no photocatalytic activity.

HO O SO3 •OH → HO +OH SO3- hv,.•OH I, III HO +O H • OH SO3- IV / OH hv

HO HO +O OH SO3- OH Aliphatic fragments + Organic acids SO3H OH +

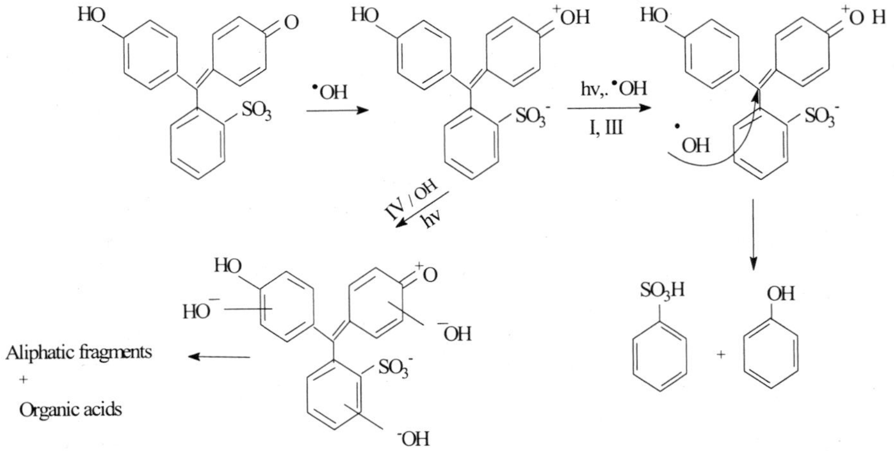

Scheme 1 : Mechanism for the photocatalyrtic degradation of PSP.

ACKNOWLEDGMENT

The authors are greatly indebted to Prof. Dr. Salah A. Hassan, Chemistry Department, Faculty of Science, Ain Shams University, Cairo, Egypt, for his valuable discussions and helpful advice during the preparation of this work.

The authors are also deeply grateful to Prof. S. Ismat Shah and Emre Yassıtepe; Ph. D. student: Physics and Astronomy, Materials Science and Engineering, University of Delaware USA, for their help in carrying out and discussing the results of XPS.

REFERENCES

[1] B.F.G. Johnson, Nanoparticles in Catalysis, Topics in Catalysis, **24,** 147- 59(2003).

[2] X. Zhang, S.A. Jenekhe, J. Perlstein, Nanoscale Size Effects on Photoconductivity of Semiconducting Polymer Thin Films, Chem. Mater., **8**, 1571- 74(1996).

[3] J.H. Fendler, Self-Assembled Nanostructured Materials, Chemical Materials, **8,**1616-24(1996).

[4] M.-C. Daniel, D. Astruc, Gold Nanoparticles: Assembly, Supramolecular Chemistry, Quantum-Size-Related Properties, and Applications toward Biology, Catalysis and Nanotechnology, Chem. Rev., **104,** 293- 346(2004).

[5] A. Corma, From Microporous to Mesoporous Molecular-Sieve Materials and Their Use in Catalysis, Chem. Rev., **97**, 2373- 419 (1997).

[6] M. Iwamoto, T. Abe, Y. Tachibana, Control Of Bandgap of Iron Oxide Through Encapsulation into SiO-Based Mesoporous Materials, J. Mol. Catal. A: Chem., **155,** 143- 53 (2000).

[7] P. Suzdalev, V.N. Buravtsev, Yu. V. Maksimov, A.A. Zharov, V.K. Imshennik, S.V. Novichikhin, V.V. Matveev, Magnetic Phase Transitions in Nanosystems: Role of Size Effects, Intercluster Interactions and Defects, J. Nanoparticle Res., **5**, 485- 95 (2003).

[8] A. Yu. Stacheev and L.M. Kustov, Effects of The Support on The Morphology and Electronic Properties of Supported Metal Clusters: Modern Concepts and Progress in 1990s, Appl. Catal. A:Gen., **188** (1-2)**,** 3-35 (1999).

[9] P. Selvam, S.E. Dapurkar, S.K. Badamali, M. Murugasan, H. Kuwano, Coexistence of Paramagnetic and Superparamagnetic Fe(III) in Mesoporous MCM-41 Silicates, Catal. Today, **68** (1), 69-74 (2001).

[10] F. Arena, G. Gatti, L. Stievano, G. Martra, S. Coluccia, F. Frusteri, L. Spadaro, A. Parmaliana, Activity Pattern Of Low-Loaded FeO_x/SiO_2 Catalysts In The Selective Oxidation of C_1 and C_3 Alkanes with Oxygen, Catal. Today, **117** (1-3), 75-79 (2006).

[11] E.-J. Shin, D.E. Miser, W.G. Chan, M.R. Hajaligol, Catalytic Cracking of Catechols and Hydroquinones in The Presence of Nano-Particle Iron Oxide, Appl. Catal. B: Environ., **61,** 79-89 (2005).

[12] H.H. Kung, M.C. Kung, Nanotechnology: Applications and Potentials for Heterogeneous Catalysis, Catal. Today, **97**, 219- 24(2004).

[13] J.L. Garcia, A. Lopez, F.T. Lazaro, C. Martinez, A. Corma, Iron Oxide Particles in Large Pore Zeolites J. Magn. and Magntic Mater., **157–158**, 272-73 (1996).

[14] G.A. Ozin, C. Gil, Intrazeolite Organometallics and Coordination Complexes: Internal versus External Confinement of Metal Guests, Chem. Rev., **89,** 1749-64 (1989).

[15] P. Behrens and G.D. Stucky, Ordered Molecular Arrays as Templates: A New Approach to The Synthesis of Mesoporous Materials, Angewandte Chem. Int. Ed., **32** (5), 696-99 (1993).

[16] A. Taguchi, F. Schüth, Ordered Mesoporous Materials in Catalysis, Micropor. Mesopor. Mater., **77 (1),** 1- 45(2005).

[17] J. Shi, Z. Hua and L. Zhang, Nanocomposites from Ordered Mesoporous Materials, J. Mater. Chem., **14**, 795- 806 (2004).

[18] P. Kustowski, L. Chmielarz, R. Dziembaj, P. Cool, E.F. Vansant, Modification of MCM-48-,

SBA-15-, MCF-, and MSU-type Mesoporous Silicas with Transition Metal Oxides Using the Molecular Designed Dispersion Method, J. Phys. Chem. B, **109**, 11552-58 (2005).

[19] P. Van Der Voort, M.B. Mitchell, E.F. Vansant and M.G. White, The Uses of Polynuclear Metal Complexes to Develop Designed Dispersions of Supported Metal Oxides: Part I. Synthesis and Characterization, Interf. Sci., **5**, 169-97 (1997).

[20] S.W. Kim, S.U. Son, T. Lee S.I. Hyeon and Y.K. Chung, Cobalt on Mesoporous Silica: The First Heterogeneous Pauson–Khand Catalyst, J. Am. Chem Soc, 122, 1550-51 (2000).

[21] C. Wu, T Bein, Conducting Polyaniline Filaments in a Mesoporous Channel Host, Science, **264**, 1757-59 (1994).

[22] Y. Plyuto, J. Berquier, C. Jacquiod and C. Ricolleau, Ag Nanoparticles Synthesised in Template-Structured Mesoporous Silica Films on a Glass Substrate, Chem. Commun.1653- (1999).

[23] G.D. Stucky, J.E. MacDougall, Quantum Confinement and Host/Guest Chemistry: Probing A New Dimension, Science **247**, 669-71 (1990).

[24] Y. Li, Z. Feng, Y. Lian, K. Sun, L. Zhang, G. Jia, Q. Yang and C. Li, Direct Synthesis of Highly Ordered Fe-SBA-15 Mesoporous Materials under Weak Acidic Conditions, Micropor Mesopor Mater, **84** (1-3), 41- 9(2005).

[25] M. Fröba, R. Köhn, G. Bouffard, Fe2O3 Nanoparticles within Mesoporous MCM-48 Silica: in Situ Formation and Characterization,Chem Mater 11, 2858-2865 (1999).

[26] I.K. Konstantinou, T.A. Albanis, TiO_2-Assisted Photocatalytic Degradation of Azo Dyes in Aqueous Solution: Kinetic and Mechanistic Investigations, Appl. Catal. B: Environ, **49**(1), 1-14 (2004).

[27] J Wang, S Uma and K.J. Klabunde, Visible Light Photocatalysis In Transition Metal Incorporated Titania-Silica Aerogels, Appl. Catal. B: Environmental, **48**(2), 151- 54(2004).

[28] F. Martinez, G. Calleja, J.A. Melero and R. Molina, Heterogeneous Photo-Fenton Degradation of Phenolic Aqueous Solutions Over Iron-Containing SBA-15 Catalyst Appl. Catal. B: Enviro., **60**(3-4), 181- 90(2005).

[29] H. Takayuki, N. Masanori, I. Komasawa., Thiol-Mediated Incorporation of Cds Nanoparticles From Reverse Micellar System into Zn-Doped SBA-15 Mesoporous Silica And Their Photocatalytic Properties, J. of Colloid and Interface Science, **268**(2), 394-99 (2003).

[30] B. Sun, E.P. Reddy and P.G. Smirniotis, TiO_2-Loaded Cr-Modified Molecular Sieves for 4-Chlorophenol Photodegradation under Visible Light, J. of Catal., **237**(2), 388-99 (2006).

[31] H.M. Ding, H. Sun, Y.K. Shan. Preparation and Characterization of Mesoporous SBA-15 Supported Dye-Sensitized TiO_2Photocatalyst, J. of Photochemistry and Photobiology A: Chemistry **169**(1), 101-107 (2005).

[32] D. Zhao, J. Feng, Q. Huo, N. Melosh, G. Fredrickson, B. Chmelka, G.D. Stucky, Triblock Copolymer Syntheses of Mesoporous Silica with Periodic 50 to 300 Ångstrom Pores, Science **279,** 548-52 (1998).

[33] F. Martinz, Y. Han, G. Stucky, J. Stotelo, G. Ovejero, J. Melero, Synthesis and Characterization of Iron-Containing SBA-15 Mesoporous Silica, Stud Surf Sci Catal., **142**, 1109-16 (2002).

[34] A. Tuel, S. Gontier, Synthesis and Characterization of Trivalent Metal Containing Mesoporous Silicas Obtained by a Neutral Templating Route, Chem. Mater., **8**, 114-122 (1996).

[35] Sing, K. S. W., Everett, D. H., Haul, R. A. W., Moscou, L.,Pierotti, R. A., Rouquerol, Reporting Physisorption Data for Gas/Solid Systems with Special Reference to The Determination of Surface Area and Porosity, Pure and Applied Chemistry, **57,** 603-18 (1985).

[36] T. Tsoncheva, J. Rosenholm, C.V. Teixeira, M. Dimitrov, M. Linden, C. Minchev, Preparation, Characterization and Catalytic Behavior in Methanol Decomposition of Nanosized Iron Oxide

Particles within Large Pore Ordered Mesoporous Silica Micropor. Mesopor. Mater., **89,** 209-18 (2006).
[37] T. Tsoncheva, J. Rosenholm, M. Linden, L. Ivanova, C. Minchev, Iron and Copper Oxide Modified SBA-15 Materials As Catalysts in Methanol Decomposition: Effect of Copolymer Template Removal, Appl. Catal. A: Gen., **318**, 234-43 (2007).
[38] L. Vradman, M.V. Landau, D. Kantorovich,Y.Koltypin, A. Gedaken, Evaluation of Metal Oxide Phase Assembling Mode inside the Nanotubular Pores of Mesostructured Silica, Micropor. Mesopor. Mater. **79,** 307–18 (2005).
[39] X. Wang, J. Jia, L. Zhao and T. Sun, Mesoporous SBA-15 Supported Iron Oxide: A Potent Catalyst for Hydrogen Sulfide Removal Water Air Soil Pollut.,**193,** 247-57 (2008).
[40] S. Zheng, L. Gao and J. K. Gao, Synthesis and Characterization of Copper(II)-Phenanthroline Complex Grafted Organic Groups Modified MCM-41, Materials Chemistry and Physics, **71**, 174- 78(2001).
[41]M. Yamane, In Sol-gel Technology for Thin Films. New Jersey: Noyes, p. 200 (1989)
[42] L. Lan, G. Gnappi and A. Montenero, Infrared Study of EPOXS-TEOS-TPOT gels. J. of Materials Science, **28,** 2119–2123 (1993).
[43] B. Notari, Microporous Crystalline Titaninum Silicate, Adv. Catal., **41**, 253-334 (1999).
[44] E. Fois, A. Gamba, G. Tabacchi, S. Coluccia and Gand Martra, Ab initio Study of Defect Sites at The Inner Surfaces of Mesoporous Silicas, J. Phys.Chem. B,107, 10767-10772 (2003).
[45] B. Tian, X. Liu, Ch. Yu, F. Gao, Q. Luo, S. Xie, B. Tu and D. Zhao, Microwave Assisted Template Removal of Siliceous Porous Materials, Chem. Commun., 1186-87 (2002).
[46] Y. M. Wang, Z.Y. Wu and J. H. Zhu, Surface, Surface Functionalization of SBA-15 by the Solvent-Free Method, J. of Solid State Chemistry, **177,** 3815-3823 (2004).
[47] Y. Abe, T. Gubji, Y. Kimata, M. Kuramata, A. Kosgoz and T. Misono, Preparation of Polymetalloxanes as a Precursor for Oxide Ceramics, J. Non-Cryst Solids, **121,** 23- 45 (1990).
[48] Z. Liu, R.J. Davis, Investigation of the structure of microporous Ti-Si mixed oxides by x-ray, UV reflectance, FT-Raman, and FT-IR spectroscopies J. Phys. Chem., **98,** 1253- 61 (1994).
[49] S. Srinirasan, A.K. Datye, M. Hampden, I.E. Smith, G. Wachs, J.M. Deo, A. Jehng, M. Truek and C.H.F. Peden, The Formation of Titanium Oxide Monolayer Coatings on Silica Surfaces J. Catal., **131** (1991) 260-75.
[50] B.L. Newalkar and S. Komarneni, Control Over Microporosity of Ordered Microporous Mesoporous Silica SBA-15 Framework under Microwave-hydrothermal Conditions: effect of salt addition , Chem. Mater., **13,** 4573- 79 (2001).
[51] L. Wang, A. Kong, B. Chen, H. Ding, Y. Shan, M. He, Direct Synthesis, Characterization of Cu-SBA-15 and Its High Catalytic Activity in Hydroxylation of Phenol By H_2O_2, J. of Molecular Catalysis A: Chemical, **230,** 143-**50** (2005).
[52] C.- M. Yang, B. Zibrowius, W. Schmidt, and F. Schüth, Materials, Stepwise Removal of The Copolymer Template Frommesopores And Micropores In SBA-15, **16**, 2918-25 (2004).
[53] D. Kantorovieh, L. Haviv, L. Vradman a and M. V. Landau, Behaviour of NiO and Ni0 phases at high loadings in SBA-15 and SBA-16 mesoporous silica matrices Studies in Surface Science and Catalysis 156 (Nanoporous Materials IV), : 147-54 (2005)
[54] K.J. Kim, D.W. Moon, S.K. Lee and K.H. Jung, Formation of a highly oriented FeO thin film by phase transition of Fe3O4 and Fe nanocrystallines Thin Solid Films **360**, 118- 21(2000).
[55] B. M. Weckhuysen, D.Wang, M.P. Rosynek and J.H. Lunsford, Conversion of Methane to Benzene over Transition Metal Ion ZSM-5 Zeolites: II. Catalyst Characterization by X-Ray Photoelectron Spectroscopy, J. Catal., **175,** 347-351 (1998).
[56] A.P. Grosvenor, B.A. Kobe, M.C.Biesigner, N.S. Mclntyre, Investigation of multiplet splitting of Fe 2p XPS spectra and bonding in iron compounds Surf Interface Anal., **36,** 1564-74

(2004).
[57] T. Y. Wei., Y. Y.Wang. and Ch. ChWan, Photocatalytic Oxidation of Phenol in the Presence of Hydrogen Peroxide and Titaninum Dioxide Powders, J. of Photochem. Photoboil. A: Chem, **55**, 115-20 (1990).
[58] Z. Zhang, C.-C.Wang, R. Zakaria and J.Y. Ying, Role of Particle Size in Nanocrystalline TiO_2-based Photocatalysts, J. Phys. Chem. B, **102** 10871- 78(1998).
[59] M. Zhoua, J. Yu and B. Chenga J. Hazard. Mater., Effects of Fe-Doping on the Photocatalytic Activity of Mesoporous TiO_2 Powders Prepared by an Ultrasonic Method, Mater. B **137**, 1838-47 (2006).
[60] W.-C. Hung, S.-H. Fu, J.-J. Tseng, H. Chu and T.-H. Ko, Study on Photocatalytic Degradation of Gaseous Dichloromethaneusing Pure and Iron Ion-Doped TiO_2 Prepared By The Sol–Gel Method, Chemosphere, **66**, 2142- 51(2007).
[61] J. Zhu, W. Zheng, B. He, J. Zhang and M. Anpo, Characterization of Fe–TiO_2 Photocatalysts Synthesized by Hydrothermal Method and Their Photocatalytic Reactivity for Photodegradation of XRG Dye Diluted in Water, J. Mol. Catal. A: **216** (1), 35-43 (2004).

PATTERNING BY FOCUSED ION BEAM ASSISTED ANODIZATION

J. Zhao, K. Lu, B. Chen, Z. Tian
Department of Materials Science and Engineering
Virginia Polytechnic Institute and State University
Blacksburg, Virginia, USA

ABSTRACT

Nano-patterning by anodization has attracted wide attention because of its potential to produce templates for unique nanostructure formation. However, anodization has a very narrow window to produce nanopore arrays and cannot produce highly ordered nanopore arrays in a large area. Focused ion beam, an advanced lithographic technique, can be used to make initiation sites for ordered nanopore arrays within a short time. In this study, highly ordered, shallow, hexagonal nanopore arrays are first made on high purity aluminum by focused ion beam lithography. These shallow pore arrays are then used as initiation sites during anodization to form pores in conjunction with the intrinsic pores to be formed from anodization. Ordered nanopore arrays can be more easily obtained by combining focused ion beam lithography and anodization. The nanopore diameter and pore-to-pore distance are dictated by anodization condition.

INTRODUCTION

Nanopattern formation by top-down approach progresses from the macro-scale to the nanoscale by starting with bulk materials and subtractively creating nano-features. But this process is expensive and the surface area that can be patterned is limited. For applications that require large response areas such as solar cells, photoluminescence, and sensors, these problems need to be addressed in order to create well-controlled patterns with greater sophistication.

In recent years, anodization has attracted considerable attention for fabrication of hexagonally-packed nanopatterns.[1] The most well-known process for creating self-ordered nanopores is two-step anodization.[2,3] Self-ordered nanopore arrays of different pore sizes and pore intervals are reported for oxalic acid, sulfuric acid, and phosphoric acid.[4,5] Anodization of aluminum is a highly desired top-down technique to create hexagonally self-organized nanopores in a parallel fashion. The method is inexpensive and can be practiced in any standard lab.[2,6-9] Pore size and depth can be varied by changing applied voltage, anodization time, and electrolytes.[10-12] At the Al/Al_2O_3 interface, an Al_2O_3 barrier film forms and the electric field determines the film growth rate. At the Al_2O_3/electrolyte interface, Al_2O_3 dissolves and the stress field, assisted by local heating, determines Al_2O_3 dissolution rate.[13] The volume change due to Al_2O_3 formation and thermal expansion introduces stress within the barrier layer and self-organizes the pore pattern. High pore areal densities of 10^{11} pores/cm^2 can be made.

However, there are several deficiencies that need to be addressed in order to advance the field. The first deficiency is nanopattern non-uniformity. Even though much work has been reported and even commercial products are available, most anodization samples are only membrane quality instead of template quality for a lack of uniform pore size and pore pattern. This deficiency is often carried into templated materials. The second deficiency is lack of large area nanopatterns. The naturally occurring defect-free area is limited to several square microns with defects concentrating on the boundaries between domains.[14,15]

Indentation was attempted to pre-texture aluminum surface to address such issues.[16] However, the pre-textured pore size, shape, and spacing cannot be accurately controlled. Also, nano-indentation is a time-consuming process for surface patterning. In addition to chemically cleaning the aluminum surface, electropolishing has been deliberately carried out to produce striped and hexagonal patterns. However, the pre-textured patterns reveal no connection with the self-organized pore patterns after anodization; the pre-textured pattern spacing shows no connection with the anodized pore spacing.[17,18] Nevertheless, electropolishing improves pore self-organization tendency. Mechanical indentation shows the potential to direct pore site and spacing. Our hypothesis is that pre-texturing can dictate nanopattern formation but there is a threshold depth for the pre-textured sites to be effective. This value is unknown but is believed to be around 5 nm. The concaves produced by electropolishing are too shallow while the pores produced by indentation are unnecessarily deep (11-23 nm) with a huge variation (>10 nm).

Focused ion beam (FIB) patterning can be used to make initiation sites for ordered nanopore arrays within a short time. Hierarchical pattern formation by FIB has been the subject of our recent study.[19] The pre-textured initiation sites are 5-8 nm deep. This is mainly achieved by the combination of low ion dose (1.5 pA current) and simultaneous electron beam imaging (1.5 nm resolution). No other technique can target the sites as precisely as FIB.[20] In comparison to nano-indentation, dual beam FIB offers much improved pore initiation site pattern and size control by eliminating mechanical contact, load variation, tip wear, and the fundamental limit of tip size. The 2D pattern densities achieved by FIB and electron beam lithography are comparable, with site to site spacing at ~10 nm.[21] However, FIB can work on almost any materials with less proximity effect while electron beam lithography is mainly effective for soft materials.[22-24] FIB pre-texturing is also extremely fast.

In this study, we have used dual beam FIB maskless patterning to initiate the anodization sites. Because of the availability of the electron beam for imaging, the ions are only present during site initiation and the pre-textured patterns are not affected by the ion sputtering that follows. The pre-textured patterns have much improved size accuracy because of precise ion dose control and ≤ 5 nm spot size. In comparison to the square microns of uniform surface area offered by pure anodization, FIB can be programmed to uniformly pre-texture larger surface areas. FIB pre-texturing can overcome the domain boundary problem by the overriding role of pre-textured sites in dictating nanopattern development.

EXPERIMENTAL PROCEDURE

Prior to anodization, high purity aluminum sheets (99.999%, Goodfellow Corporation, Oakdale, PA) were cut into 8 mm×18 mm×0.3 mm size pieces. The small pieces were mechanically polished with 600 mesh and 1200 mesh polishing papers followed by a 5 micron alumina suspension. Then the aluminum pieces were annealed in a high purity argon atmosphere at 500°C for 24 hrs in order to recrystallize aluminum and remove mechanical stress. The annealed aluminum pieces were immersed in a 0.5 wt% NaOH solution for 5 min, rinsed with deionized water, degreased in acetone, and washed in deionized water to remove the oxidized surface layer. After that, the aluminum pieces were electropolished in a 1:4 mixture of perchloric acid (60%-62%):ethanol (95%) (volume ratio) under a constant voltage of 12V with the bath temperature at 10°C for 30 seconds. This process was intended to reduce the roughness of the aluminum surface. Finally these aluminum substrates were rinsed with deionized water and kept in ethanol.

For the samples that were anodized directly, the anodization electrolyte was phosphoric acid or sulfuric acid. For the former, the anodization condition was 0.1 M H_3PO_4 at 150, 165, and 190V for 10 min at 0°C or 0.4 M H_3PO_4 at 150 V for 10 min at 0°C. For the latter, the anodization condition was 0.4 M H_2SO_4 at 25V for different time at 0°C.

A dual beam FIB (FEI Helios 600 NanoLab, HillsBoro, OR) was employed to create different patterns on the aluminum surface to guide the anodization. The instrument was composed of a sub-nanometer resolution field emission scanning electron microscope (SEM) and a field emission scanning Ga^+ beam column. The aluminum samples can be moved by 150 mm distance along X and Y axes and tilted from -5 to 60° by a high precision specimen goniometer. The Ga^+ source had a continuously adjustable energy range from 0.5 kV to 30 kV, and an ion current between 1.5 pA and 21 nA. The patterning beam spot size and resolution varied with the ion beam current and voltage and can be tuned to as small as 5 nm. The FIB instrument also had a high resolution, 24 bit digital patterning engine capable of simultaneous patterning and imaging. All the FIB patterns created in this work were 2050 nm×2050 nm in size with 4100×4100 pixels. All the pore diameters were 50 nm. Four different interpore distances were used: 100 nm, 200 nm, 300 nm, and 400 nm.

The FIB pre-patterned aluminum substrates were also anodized in phosphoric acid or sulfuric acid. The anodization in the phosphoric acid electrolyte was performed at 0.2 M concentration with 140 V voltage for 10 minutes at 0°C, while in the sulfuric acid electrolyte it was conducted at 0.3 M with a constant voltage of 25 V for 10 minutes at 0°C. After the anodization in the acid solutions, the aluminum samples were rinsed with deionized water.

The surface morphologies of the anodized samples with and without the FIB patterning were investigated by SEM (Quanta 600 FEG, FEI Company, Hillsboro, OR).

RESULTS AND DISCUSSION

Without electropolishing, the aluminum surface is rough and nanopores cannot organize into hexagonal patterns. With electropolishing, when the aluminum samples are anodized in phosphoric acid or sulfuric acid, naturally self-organized pore arrays should form with a certain diameter and interpore distance, which are determined by the acid concentration and voltage. The SEM images of the aluminum samples after the electropolishing are shown in Figure 1. The aluminum surface is smooth with some dust particles on it (Figure 1(a)). At high magnification (Figure 1(b)), however, some shallow and irregular concaves can be observed. The inter-distance of the concaves on the surface is 15~20 nm, with only several nanometers depth. Some studies indicate that when the aluminum surface roughness is reduced to several nanometers, the pattern on the electropolished aluminum surface cannot be maintained during anodization.[1] Therefore, these shallow and irregular concaves are not deep enough to guide the development of the pores during anodization. The main function that the electropolishing plays is to provide a smooth aluminum surface for the anodization process.

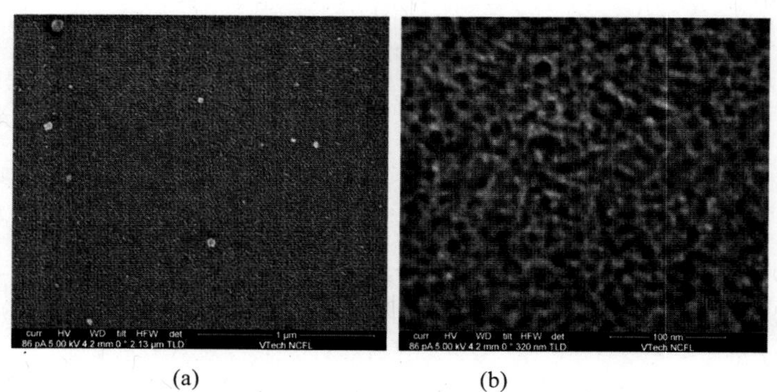

(a) (b)

Figure 1. SEM images of electropolished aluminum surface in a mixture of perchloric acid and ethanol: (a) low magnification, (b) high magnification.

Highly ordered pore arrangements after the FIB patterning process can be seen in Figure 2 with the pore-to-pore distance at 100 nm, 200 nm, 300 nm, and 400 nm respectively in 2050 nm×2050 nm areas. All the nanopore diameters in Figure 2 are 50 nm with the pore depth at 5-8 nm. Since the FIB patterning beam spot size and resolution can be tuned to as small as 5 nm, the nanopores have roughly the same size and depth, and are in good hexagonal arrangement. At the same time the region without the FIB patterning remains unchanged. Based on the literature, the hexagonally close-packed concaves deep than 5 nm can effectively guide the growth of alumina nanopores during anodization.[2] By designing patterns with different inter-pore distances on the aluminum surface, the effect of the FIB patterning on the anodization pattern formation can be examined.

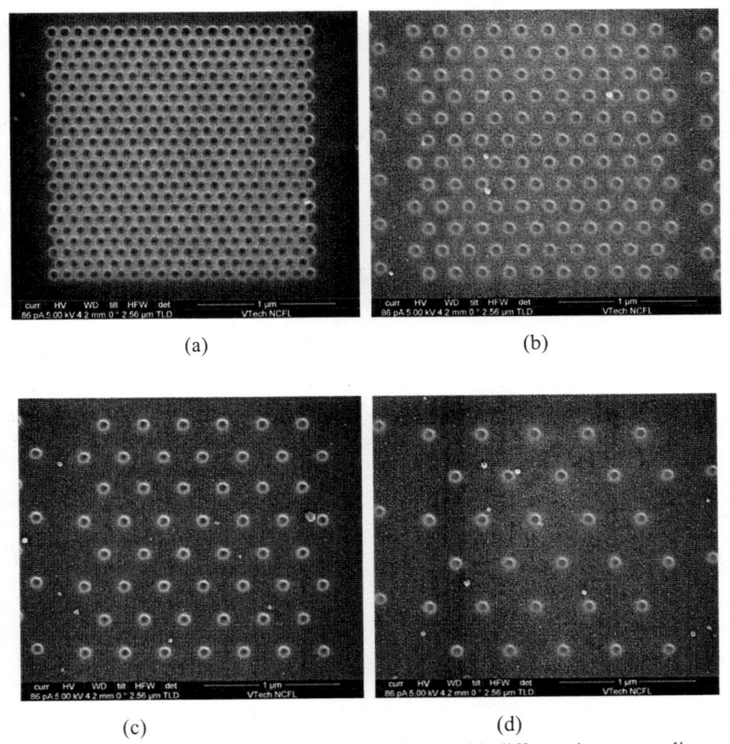

Figure 2. SEM images of FIB patterned aluminum surfaces with different inter-pore distance: (a) 100 nm, (b) 200 nm, (c) 300 nm, and (d) 400 nm.

Figure 3 shows the SEM images of the anodized aluminum substrates by the anodization process in the phosphoric acid electrolyte. Figures 3(a)-(c) have the same phosphoric acid concentration, 0.1 M, while the anodization voltage varies from 150 V, to 165 V, and to 190 V. Figures 3(a) and (d) have the same anodization voltage, 150V, but different phosphoric acid electrolyte concentration. For Figure 3(a), the electrolyte concentration is 0.1 M. For Figure 3(d), the electrolyte concentration is 0.4 M. The anodization time is 3 min for all the samples.

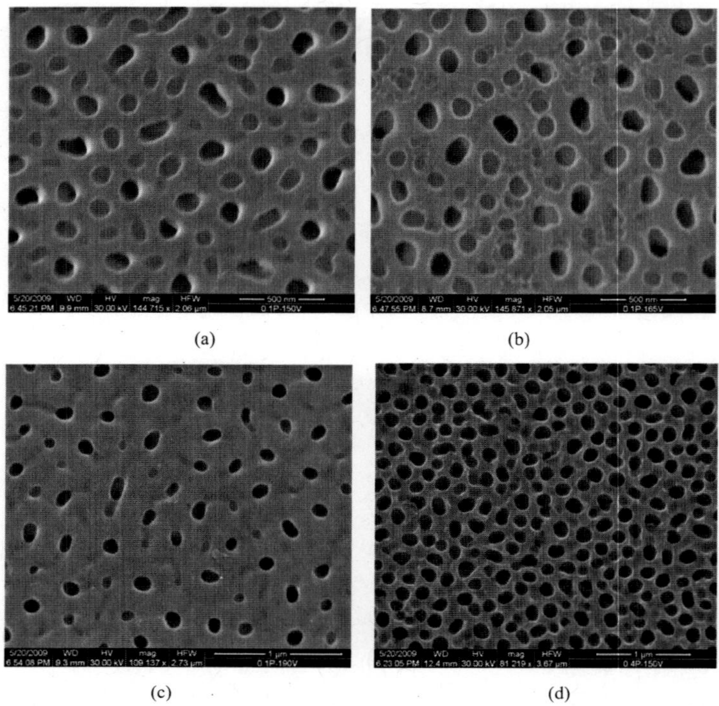

Figure 3. SEM images of anodized aluminum substrates in phosphoric acid electrolyte: (a) 150V, (b) 165V, (c) 190V, and (d) 150V. (a) to (c) has 0.1 M phosphoric acid concentration. (d) has 0.4 M phosphoric acid concentration.

As shown in Figure 3, nanopores appear on all the aluminum sample surfaces after 3 min of anodization. Since the anodization time is relatively short, the images may just reveal the initial pore growth trend for the aluminum substrate in the phosphoric acid environment. From Figure 3(a) to Figure 3(c), the inter-pore distance increases with increasing anodization voltage. It is around 240 nm for Figure 3(a). The pores in Figure 3(b) are quite irregular. There are many small pores appear between the large pores. The formation of those small pores may be due to a slight change of voltage, which is 15 V higher than that in Figure 3(a). The small pores on the surface can influence the growth of the Al_2O_3 pore pattern, leading to an irregular pore distribution as the anodization continues. If we ignore the small pores, the inter-pore distance for the large pores in Figure 3(b) varies from 180 nm to 260 nm. When the anodization voltage increases to 190 V (Figure 3(c)), there are fewer small pores on the sample surface and the inter-pore distance increases to around 320 nm. Besides the inter-pore distance, the three samples do not exhibit obvious change in pore size. Based on these observations, it seems that the nanopore distances formed during the first three minutes is sensitive to anodization voltage, which would eventually influence the inter-pore distance and the pore density. Future work should be conducted to understand the subsequent growth of the nanopores after long time anodization.

In order to understand the effect of the phosphoric acid concentration on anodization, the phosphoric acid concentration is increased to 0.4 M for Figure 3(d), four times that of Figure 3(a), while the anodization voltage and time stay the same. The SEM image (Figure 3(d)) shows that the pores are deeper, the pore density is higher, and the pore size is larger compared to the samples anodized in the 0.1 M phosphoric acid electrolyte. Higher phosphoric acid concentration increases the nanopore formation and growth rates. The inter-pore distance for Figure 3(d) varies from 200 nm to 250 nm.

Figure 4 shows the SEM images of the samples that were patterned by the FIB before anodization in the phosphoric acid electrolyte. For the FIB patterns, the inter-pore distances are 100 nm, 200 nm, 300 nm, and 400 nm, respectively, from Figure 4(a) to 4(d). The anodization time is 10 min in order to investigate the growth of the alumina nanopores. Except for Figure 4(a), the pore arrays are much more regular than those without the FIB patterning, as can be seen from the pores around the edges of the images. In Figure 4(b), the pore depths are different. These means the pore growth rate on the same surface is different. This different pore growth rate can just be a phenomenon for the initial pore growth. More anodization time might be needed for the pores to fully develop and this needs to be studied. Figures 4(c) and (d) have the best pore shape and arrangement. The pore diameter and depth are the most uniform. Since the four samples were anodized under the same condition, the difference in pore patterns should be from the FIB patterning step. It indicates that 300 nm and 400 nm inter-pore distance samples, which match the self-organized nanopore distance under the specific anodization condition, have the best anodization result. Compared to anodization only, FIB patterning facilitates ordered pore pattern formation and growth during the subsequent anodization process. It is highly possible that if the inter-pore distance designed by the FIB patterning is similar to that of the anodization, it could greatly regulate the pore pattern on the aluminum substrate.

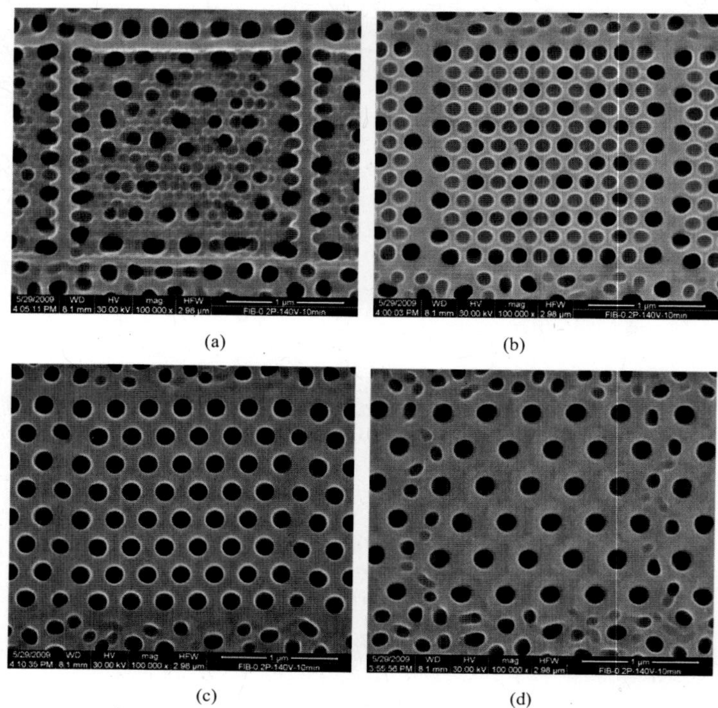

Figure 4. SEM images of the samples patterned by FIB patterning before anodization in 0.2 M phosphoric acid electrolyte at 140 V: (a) 100 nm, (b) 200 nm, (c) 300 nm, and (d) 400 nm.

In order to understand whether the FIB patterning has similar functions in other electrolyte systems, anodization of the aluminum substrate has also been carried out for the sulfuric acid electrolyte. The SEM images of the anodized samples in the sulfuric acid electrolyte are shown in Figure 5. Figure 5(a) is the sample without the FIB patterning and Figure 5(b) is the sample patterned by the FIB before anodization. After 5 minutes of anodization without the FIB patterning, pore size and arrangement are irregular. After 10 minutes of anodization with the FIB patterning, uniform pore size and pattern are obtained. The pore diameter is about 45 nm and the pore-to-pore distance is around 60 nm. However, anodization by the sulfuric acid is a rapid process (also called hard anodization). The patterns from the FIB patterning cannot be maintained. Only self-occurring pores exist. It can be observed that the aluminum surface that has been swept by the focused ion beam offers much better self-organized pore pattern after anodization. For the aluminum surface that has never been swept by the ion beam, irregular size and arrangement pores are created. This means that FIB plays a role in influencing the pore development even without purposely designed patterns. This aspect needs to be studied in detail in the future.

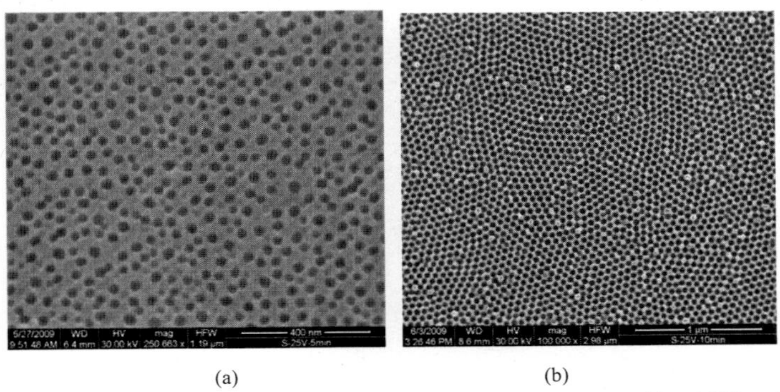

(a) (b)

Figure 5. SEM images of the anodized samples in 0.3 M sulfuric acid at 25 V and 0°C: (a) without FIB patterning, (b) with FIB patterning before anodization.

CONCLUSION

In this study, focused ion beam lithography has been used to prepare ordered shallow nanopore arrays on aluminum surface. These shallow pores are then used as initiation sites to lead the nanopore pattern formation during anodization. Anodizaiton in phosphoric acid under several different voltages creates irregular pore diameter and disordered pore arrangement. With the FIB patterning before anodization, ordered nanopore arrays with uniform pore diameter are formed. Compared to the irregular nanopore arrays formed by anodization only, the best pre-texturing effect is obtained when phosphoric acid is used as the electrolyte and the inter-pore distance designed by the FIB patterning matches that by anodization. Anodization in sulfuric acid cannot produce the pore arrays defined by the FIB patterning; but FIB sweeping of the aluminum surface leads to very uniform pore diameter and nanopore arrays for the subsequent anodization process.

ACKNOWLEDGMENT

The authors acknowledge the financial support from National Science Foundation under grant No. CMMI-0824741.

REFERENCES

[1]S. Ono, M. Saito, and H. Asoh, Self-ordering of anodic porous alumina formed in organic acid electrolytes, *Electrochim. Acta*, **51**, 827–833 (2005).

[2]H. Masuda, K. Fukuda, Ordered metal nanohole arrays made by a 2-step replication of honeycomb structures of anodic alumina, *Science*, **268**, 1466-1468 (1995).

[3]H. Masuda, K. Yasui, and K. Nishio, Fabrication of ordered arrays of multiple nanodots using anodic porous alumina as an evaporation mask, *Adv. Mater.*, **12**, 1031-1033 (2000).

[4]A. P. Li, F. Muller, A. Birner, K. Nielsch, and U. Gosele, Hexagonal pore arrays with a 50-420 nm interpore distance formed by self-organization in anodic alumina, *J. Appl. Phys.*, **84**, 6023-6026 (1998).

[5]H. Masuda, K. Yada, and A. Osaka, Self-ordering of cell configuration of anodic porous alumina with

large-size pores in phosphoric acid solution, *Jpn. J. Appl. Phys.* **37**, L1340-L1342 (1998).
[6]C. R. Martin, Nanomaterials - a membrane-based synthetic approach, *Science*, **266**, 1961-1966 (1994).
[7]H. Chik, J. M. Xu, Nanometric superlattices: non-lithographic fabrication, materials, and prospects, *Mat. Sci. Eng. R*, **43**, 103-138 (2004).
[8]K. Nielsch, J. Choi, K. Schwirn, R. B. Wehrspohn, and U. Gosele, Self-ordering regimes of porous alumina: The 10% porosity rule, *Nano Lett.*, **2**, 677-680 (2002).
[9]S. Ono, N. Masuko, Evaluation of pore diameter of anodic porous films formed on aluminum, *Surf. Coat. Tech.*, **169**, 139-142 (2003).
[10]S. Shingubara, O. Okino, Y. Sayama, H. Sakaue, and T. Takahagi, Ordered two-dimensional nanowire array formation using self-organized nanoholes of anodically oxidized aluminum, *Jpn. J. Appl. Phys.*, **36**, 7791-7795 (1997).
[11]H. Masuda, F. Hasegwa, S. Ono, Self-ordering of cell arrangement of anodic porous alumina formed in sulfuric acid solution, *J. Electrochem. Soc.*, **144**, L127-L130 (1997).
[12]S. Z. Chu, K. Wada, S. Inoue, M. Isogai, Y. Katsuta, and A. Yasumori, Large-scale fabrication of ordered nanoporous alumina films with arbitrary pore intervals by critical-potential anodization, *J. Electrochem. Soc*, **153**, B384-B391 (2006).
[13]F. Y. Li, L. Zhang, and R. M. Metzger, On the growth of highly ordered pores in anodized aluminum oxide, *Chem. Mater.*, **10**, 2470-2480 (1998).
[14]C. H. Liu, J. A. Zapien, Y. Yao, X. M. Meng, C. S. Lee, S. S. Fan, Y. Lifshitz, and S. T. Lee, High-density, ordered ultraviolet light-emitting ZnO nanowire arrays, *Adv. Mater.*, **15**, 838-841 (2003).
[15]H. Asoh, K. Nishio, M. Nakao, T. Tamamura, and H. Masuda, Conditions for fabrication of ideally ordered anodic porous alumina using pretextured Al, *J. Electrochem. Soc.*, **148**, B152-B156 (2001).
[16]Y. Matsui, K. Nishio, and H. Masuda, Highly ordered anodic porous alumina by imprinting using Ni molds prepared from ordered array of polystyrene particles, *Jpn. J. Appl. Phys.*, **44**, 7726-7728 (2005).
[17]V. V. Yuzhakov, H. C. Chang, and A. E. Miller, Pattern formation during electropolishing, *Phys. Rev. B*, **56**, 12608-12624 (1997).
[18]M. T. Wu, I. C. Leu, and M. H. Hon, Effect of polishing pretreatment on the fabrication of ordered nanopore arrays on aluminum foils by anodization, *J. Vac. Sci. Technol. B*, **20**, 776-782 (2002).
[19]K. Lu, Hierarchical sized nanopattern formation using dual beam focused ion beam, *J Nanosci. Nanotechnol.*, **9**, 2598-2602 (2009).
[20]J. Mayer, L. A. Giannuzzi, T. Kamino, and J. Michael, TEM sample preparation and FIB-induced damage, *MRS Bull.*, **32**, 400-407 (2007).
[21]R. M. Langford, P. M. Nellen, J. Gierak, and Y. Q. Fu, Focused ion beam micro- and nanoengineering, *MRS Bull.*, **32**, 417-423 (2007).
[22]M. Catalano, A. Taurino, M. Lomascolo, A. Schertel, and A. Orchowski, Critical issues in the focused ion beam patterning of nanometric hole matrixes on GaAs based semiconducting devices, *Nanotechnology*, **17**, 1758-1762 (2006).
[23]J. Gierak, E. Bourhis, A. Madouri, M. Strassner, I. Sagnes, S. Bouchoule, M. N. Combes, D. Mailly, P. Hawkes, R. Jede, L. Bardotti, B. Prevel, A. Hannour, P. Melinon, A. Perez, J. Ferre, J. P. Jamet, A. Mougin, C. Chappert, and V. Mathet, Exploration of the ultimate patterning potential of focused ion beams, *J. Microlith. Microfab.*, **5**, 011011-1-11 (2006).
[24]Y. K. Kim, A. J. Danner, J. J. Raftery, and K. D. Choquette, Focused ion beam nanopatterning for optoelectronic device fabrication, *IEEE J. Sel. Top. Quantum Electron.* **11**, 1292-1298 (2005).

AGRICULTURAL-WASTE NANO-PARTICLE SYNTHESIS TEMPLATES FOR HYDROGEN STORAGE

William L. Bradbury, Eugene A. Olevsky
Department of Mechanical Engineering, San Diego State University
San Diego, CA, US

ABSTRACT

Agricultural-waste biomass is carbonized and chemically activated, intended for utilization as a template to tailor nano-structured adsorption materials for solid state atomic hydrogen storage systems. These natural materials exhibit well developed surface area and structural features which may provide preferential physisorption sites for molecular hydrogen gas. The production of silicon carbide nano-structures upon and within the biomass-carbon templates is attempted to provide preferential adsorption sites for atomic hydrogen vapor and potentially further strengthen the proposed adsorption bed structure without significantly reducing the specific surface area. Material characterization includes SEM, EDX, XRD, TEM and BET SSA analysis.

INTRODUCTION

Carbonization and further activation of agricultural and other renewable biomass waste materials is a promising technique for the development of high specific surface area (SSA) micro- and nano-porous materials. Recent research has shown the potential for these materials to act as naturally tailored nano-structured particle synthesis templates[1] and solid-state reversible adsorptive media for particle[2] and vapor storage systems[3]. Hundred millions of tonnes of agricultural waste are produced annually, and much of this material is harvested or destroyed through low efficiency energy capture processes.

Activated Carbon (AC) obtained from biomass sources has shown promising capacity for reversible adsorptive storage of hydrogen at cryogenic temperatures and moderate pressures. Yakovlev et. al[4] demonstrated that rice-husk can yield carbonaceous materials with SSA values of up to ~3400 m^2/g, which will adsorb ~6wt% hydrogen at -196 °C. Cheng et. al[5] produced biomass AC through carbonization of sawdust, which exhibited SSA values of ~3100 m^2/g and adsorbed ~5wt% hydrogen at -196 °C.

Many proposed adsorbents are tested in a powdered form and exhibit some hysteresis due to macroscopic changes or damage to the micro- or macro-structure, potentially reducing the efficiency of the adsorbent material system[6]. A component must then be designed for optimal resistance to loads imposed during purge and saturation cycles as well as strong vibrational forces particularly encountered in transportation applications. The transformation of activated carbon materials to silicon carbide (SiC) may provide a partial solution to the problem of durability and resistance to fatigue as well as provide additional desirable properties.

A number of novel Silicon carbide nano-structures have been produced, such as nanotubes, nanowires and nanoflowers. They synthesis and diversity of these materials is the subject of an eloquent review and discussion by Fan et. al[7]. SiC nano-materials have exhibited significant hydrogen adsorption potential in experimental[8] and computational studies[9]. A comparison of silicon carbide nanotubes (SiCNT) generated through chemical vapor deposition (CVD) templating on multi-walled carbon nanotubes and the precursor material reported a twofold enhanced gravimetric (wt%) adsorption capability of hydrogen gas on SiCNT products with less than 1/3 relative SSA[10].

It is an opened question as to whether similar enhancements to the properties of biomass materials for hydrogen storage can be realized through the production of biomass silicon carbide materials. The utilization of biomass precursors has been reported by Galvagnos et. al[11] for the production of 30-100nm spherical SiC particles from poplar char as well as by González et. al for the generation of structurally stable silicon carbide forms from beech wood[12].

The synthesis of nano-scale graphite-potassium intercalation compounds (K-GICs) during conventional potassium hydroxide (KOH) activation of carbon has been reported previously through preliminary characterization by Xue and Shen via X-ray powder diffraction (XRD)[13]. Xue et. al performed similar activation, providing electron microscopy evidence of the development of carbon nanofibers (CNF) as well as stacked carbon platelets suggesting that Co and KOH have a significant role in the catalytic growth of these particular nano-structures[14]. It was also recently proposed that natural metallic content in the original biomass composition is responsible for the formation of carbon nanofibers on activated carbon templates during CVD processing[1].

Here we report the development of high surface area carbon from specific food and oil-seed agricultural-waste biomass precursor materials. In addition, we present preliminary evidence of microstructural evolution of graphene forms, potential production of graphite intercalation compounds (GICs) and the direct synthesis of silicon carbide nano-wires on biomass-carbon nano-material templates.

PRODUCTION OF AGRICULTURAL-WASTE BIOMASS-ACTIVATED CARBON

Agricultural waste samples of flax, crambe, camelina, and lentil stalk were obtained from North Dakota State University, Williston Research Extension Center. ACS certified potassium hydroxide (KOH) was obtained from Fisher Scientific Inc. Activated carbon samples are prepared by a method similar to that described by Zhang et. al[3]. Biomass stalk is washed with deionized water and dried under vacuum. The samples are placed in Al_2O_3 crucibles and loaded into a horizontal tube furnace, heated to 500 °C and held at this temperature for 3 hours under N_2 atmosphere. The carbonized products were immersed in 25ml of ~10mol/L KOH solution for 24 hours. The impregnated carbon products were washed, separated from the solution by filtration and again loaded into a horizontal tube furnace for activation. The products were then heated to 800 °C in nitrogen atmosphere and held for 2 hours. After activation, the samples were recovered from the tube furnace, washed, and treated with a dilute HCl solution. The final products were washed until a pH value of ~7.0 was reached then vacuum dried at 100 °C for 24 hours.

Characterization of Activated Carbons

Surface area analysis is performed with a nitrogen based single point Monosorb BET adsorption system. Biomass carbon samples are purged of residual gas and impurities by heating at 250 °C for ~30 minutes. Weight-normalized BET SSA data for the samples is shown in Figure 1, where non-powdered Crambe-based biomass showed the highest SSA value of ~933 m^2/g^{-1}.

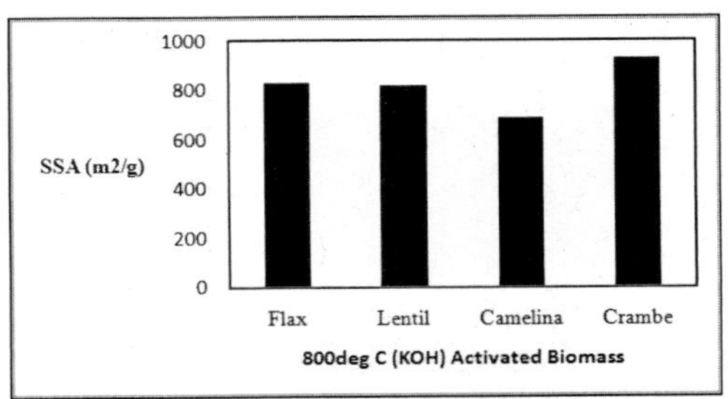

Figure 1: BET adsorption SSA measurement data for the four biomass-carbon samples.

SEM characterization was carried out with a FEI (Philips) XL-40 FEG Scanning Electron Microscope. As seen in Figure 2 (a), some pores can be discerned as ≤100nm in diameter. Many pores <1μm in diameter can be observed in both Figure 2 (a) and (b). The samples were subjected to energy-dispersive X-ray spectroscopy (EDS) which indicated the existence of residual impurities in the AC material as shown in Figure 2 (c).

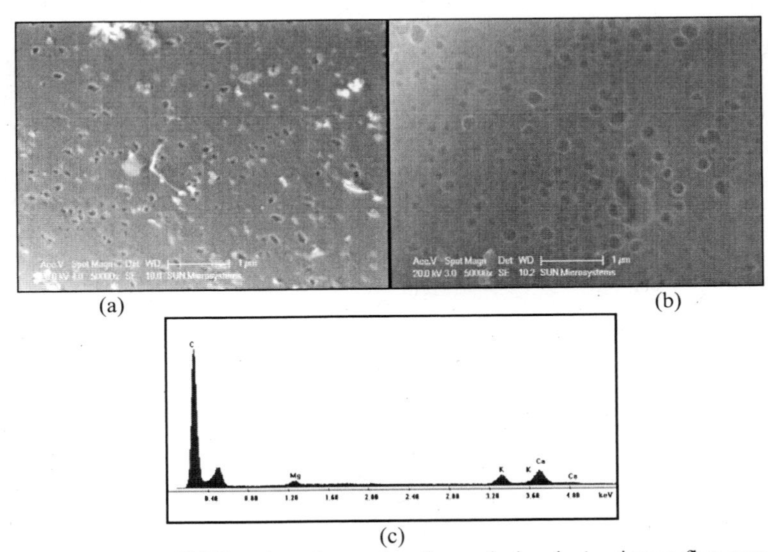

Figure 2: Nano-porous KOH activated samples, the scale bar is 1 micron; flax sample (a), crambe sample (b) and representative EDX analysis of (a) and (b) showing some impurities (c).

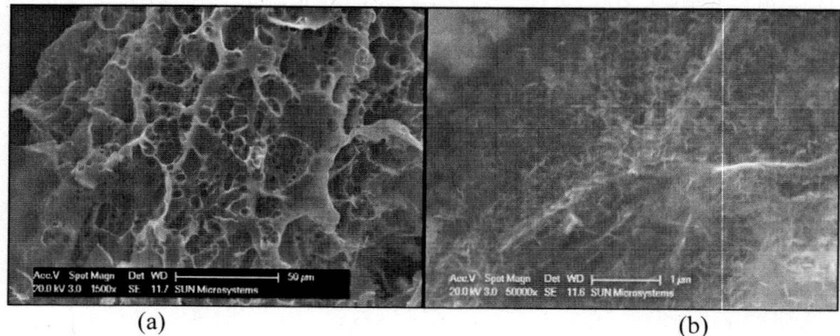

(a) (b)

Figure 3: Micro-textured KOH activated camelina sample, the scalebar is 50μm (a), lentil sample, showing nano- features, the scalebar is 1μm (b).

Figure 3 (a) represents large scale porosity, shown on the exterior of the camelina sample. Figure 3 (b) shows nano-texture resembling bundles of nanowires immobilized on the surface of the activated carbon.

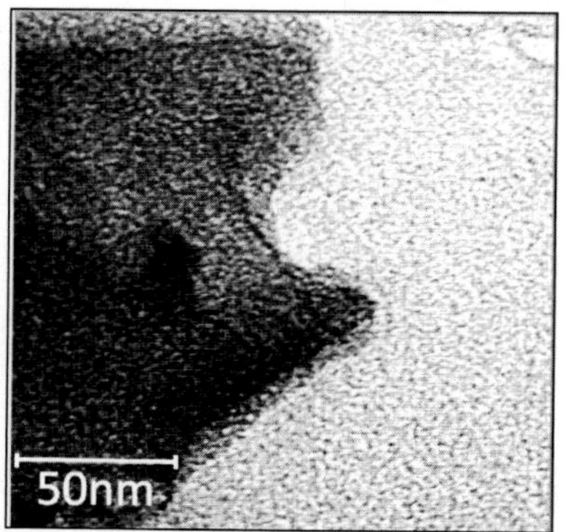

Figure 4: TEM images of activated carbons from Crambe showing nano and sub-nano-porosity.

A FEI Technai 12 Transmission Electron Microscope was utilized for analysis of nano-porosity in the biomass-activated carbon samples. A small amount of sample was lightly ground in a mortar and pestle, dispersed in ethanol and mounted by simple evaporation on a Formvar coated copper TEM grid. The nano-structure of these materials is porous and inhomogeneous. Pore diameters in the range of 2.5 - 0.5 nm can be seen in Figure 4.

X-ray diffraction (XRD) analysis is performed with a Philips X'Pert Pro Cu-K_α system. The samples exhibit a broad peak as seen in Figure 5, which corresponds to a characteristic amorphous carbons peak at ~23.6 ° two theta[3].

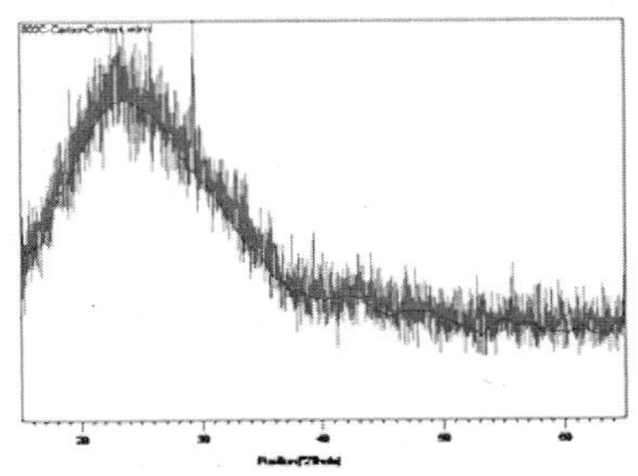

Figure 5: XRD spectra of the washed activated bio-mass carbon sample.

Characterization of Graphite-Intercalation Compounds or Nano-Flowers

The resulting microstructure shows surface decoration seemingly composed of arranged stacked carbon sheets, resembling nano-flowers particularly similar in shape and dimension to those reported by Kang et. al[15]. These novel forms are proposed as graphite-intercalation compounds (GICs) as documented in the following discussion.

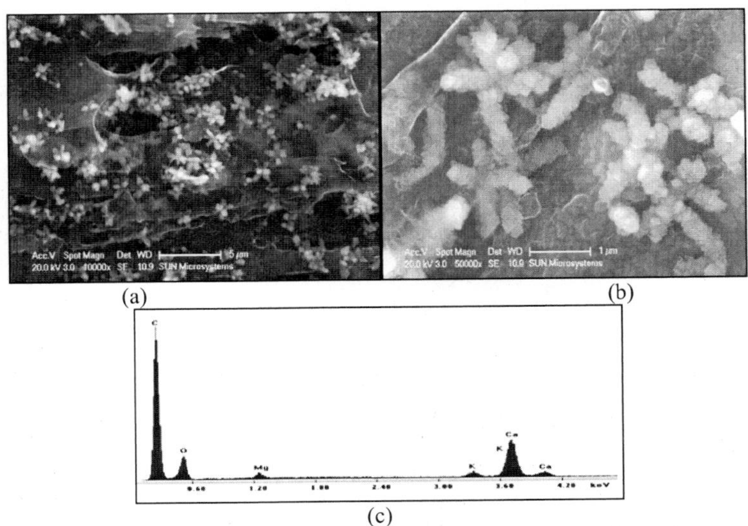

Figure 6: Macroscopic view of a field of nano-clovers and pillared graphene sheets, the scale bar is 5μm (a), Zoomed in view of various nano-structural features, pillared graphene sheets being the most prominent (b) EDX spectra of (b) at 20KeV and 50K magnification (c).

Figure 6 (a) represents a macroscopic view of the general density of these particular structures on the surface of the flax material. The morphology and density does vary widely in some localities, however those depicted in Figure 6 (a) and (b) are the most developed and repeated observed on this sample. As seen from Figure 6 (c), the sample contains calcium and oxygen, which may be calcite or calcium carbonate, as well as residual potassium.

The obtained nano-structures are found primarily on the surface of the flax carbon source, and exhibit moderate organization and a fair degree of reoccurrence. Carbon containing vapor, liquid tars or bio-oils released from the biomaterial during activation may interact with these metallic particles and initiate catalytic growth of nanowires. The growth mechanism may be driven by a vapor-liquid-solid (VLS) reaction process[16], vapor-solid (V-S) mechanisms[17], or potentially by a tip growth model[18].

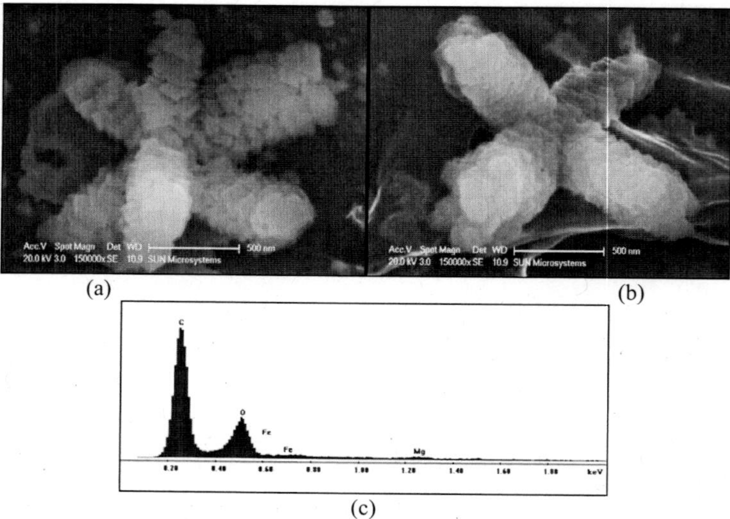

(a) (b)

(c)

Figure 7: Pillared graphene sheets or possibly GICs (a) nano-flower morphology showing 3-D structure (b), EDX spectra of (b) at 5KeV and 150K magnification (c).

A higher magnification view of the synthesized GICs or nano-flowers is shown in Figure 7 (a) with visible distinct edges of carbon or graphene sheets. As seen in Figure 7 (b), these particular microstructures may be best described as a cluster of pillared sheets catalyzed from a common source and growing in quasi-perpendicular directions. An EDX spectrum of Figure 7 (b) is shown in Figure 7 (c) indicating the relatively high purity as well as naturally occurring magnesium and iron elements.

The platelet structure may be explained by the presence of potassium or calcium during the catalysis reaction[19]. The interaction of K or Ca and C may result in intercalation of forming carbon nanowires, effectively creating the disjointed or stacked appearance[20] observed by SEM.

MICROWAVE SYNTHESIS OF SILICON CARBIDE NANO-WIRES

In attempt to achieve improved specific surface area and optimization of binding energy, the production of silicon carbide material on the as prepared biomass-activated carbon templates is performed following a technique similar to those described by G. Wei et. al[21] and J. Wei et. al[22]. Crystalline/amorphous 99.9985% purity silicon powder is obtained from Alfa Aesar Corporation. The Si powder is mechanically mixed in a 3:1 molar ratio of Crambe-based activated biomass carbon products, loaded into an alumina crucible and placed in the center of a CPI Autowave 2450MHz Microwave Generator and Applicator device chamber.

The microwave chamber is evacuated and purged with Argon gas, and a flow of ~20 ml min^{-1} is maintained over the course of the experiment. The sample temperature is monitored by a two-color infrared thermometer. The temperature is raised at ~50 °Cmin^{-1} to 1200-1400 °C, held at the peak temperature for 20 minutes and allowed to cool rapidly to room temperature.

Characterization of Microwave Synthesized SiCNW

The products of microwave synthesis show a light blue or violet textured surface coating upon removal from the microwave furnace. As seen in Figure 8 (a), microstructural features which were not part of the original activated carbon material appear, with residual silicon particles also observed. The obtained nanowires are found primarily on the surface of the biomass carbon source, and do not show a substantial degree of organization.

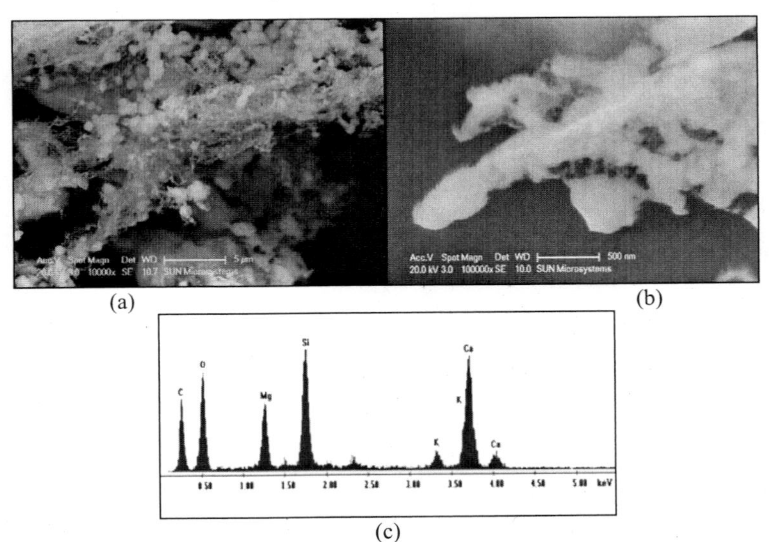

Figure 8: Macroscopic view of as-synthesized SiC nanowire product the scalebar is 5μm (a), web-like SiC nanowire feature, the scalebar is 500nm (b) and EDX spectra of (a) unpurified SiC products (c).

Figure 8 (b) shows a larger central nano/micro-wire with branching effects as well as some nano-scale webbing/netting connecting and filling in gaps between larger branches. Figure 8 (c) reveals the large amount of residual impurities still present in the sample. A relatively low percentage of carbon is seen from EDX analysis, which may be a result of carbon oxide evolution due to potential presence of oxygen in the reaction chamber or reagent gas.

CONVENTIONAL SYNTHESIS OF SILICON CARBIDE NANO-WIRES

Additionally, Si powder is ultrasonically mixed in a 3:1 molar ratio with Crambe-based activated biomass carbon products, in ethanol, dried and placed in a horizontal alumina flow through tube in a conventional tube furnace. Argon is admitted and allowed to purge the conventional tube furnace chamber and comparative flow is maintained at ~20 ml min^{-1} for the length of the experiment. The furnace is heated to 1600°C and held at this temperature for 2 hours. The temperature is monitored by an alumina sheathed thermocouple.

Characterization of Conventionally Synthesized SiCNW

The products of conventional synthesis show a dark and light green color upon removal from the tube furnace. Figure 9 (a) shows a large scale image of the conventional synthesis SiC products. The nanowires protrude from large silicon particles seen here partially embedded within the linear end of the biomass-carbon microstructure.

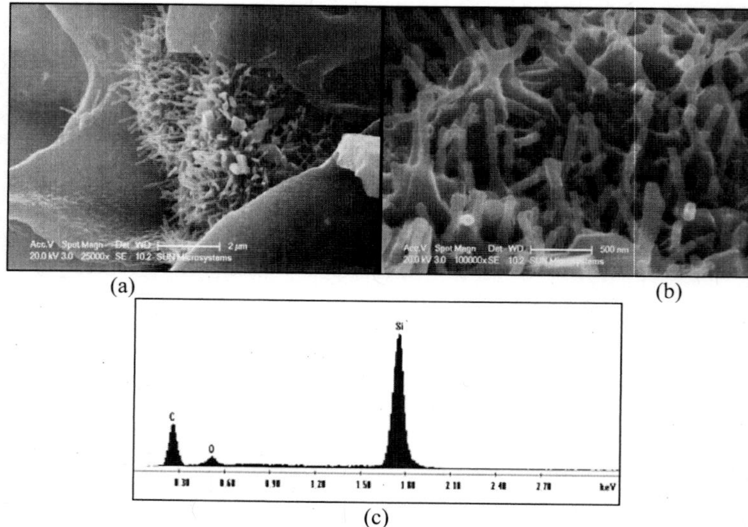

(c)

Figure 9: Macroscopic view of conventionally synthesized SiC nanowire product protruding from silicon particles, the scalebar is 2µm (a), Field of SiC nanowires where the scalebar is 500nm (b) and EDX spectra of (b) revealing relatively high purity and unreacted silicon (c).

Relatively high gas flow rates during this preliminary experiment may have resulted in turbulence through the synthesis bed, inhibiting nanowire growth by disturbance of the developed reactive layer of silicon and carbon based vapors at the surface of the biomass AC. However, local conditions within the carbon microstructure provided isolation from excessive flow and adequate conditions for SiC nanowire growth as seen in Figure 9 (b). Chemical analysis of the produced microstructure is shown in Figure 9 (c), revealing relatively high purity compared with microwave and activation experiments with expected residual silicon content.

(a) (b)

Figure 10: Triplet of SiC nanowires protruding from a small silicon particle, the scalebar is 500nm (a), relatively longer SiC nanowires, where the scalebar is 2μm (b).

Most of the nanowires observed from this conventional synthesis experiment were <500nm in length, however it is worth noting that several nanowires of much greater length were also observed, as depicted in Figure 10. Where the nano-structures shown in Figure 10 (a) are ~1μm in length, and those shown in Figure 10 (b) are ~3-5μm in length. Through manipulation and optimization of experimental conditions, favorable nanowire placement, enhancement of growth and uniformity should be relatively easily obtained. The sample was baked at 250 °C and subjected to BET SSA analysis resulting in a SSA value of $\sim 104 m^2 g^{-1}$.

For further confirmation of elemental composition, the sample is ground in an agate mortar with Isopropanol and subjected to XRD analysis as seen in Figure 11. The spectrum shows relatively high purity silicon carbide products with some residual carbon still present in the resulting semi-monolithic powder, represented by the typical broad characteristic peak.

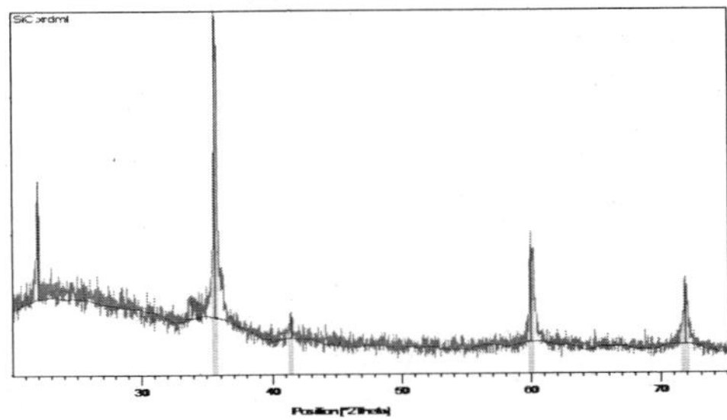

Figure 11: XRD analysis of conventionally synthesized SiC nanowire-biomass carbon products.

CONCLUSIONS

This work demonstrates the feasibility of utilizing char from biomass agricultural-waste materials for the production of high surface area activated carbon templates on which inorganic ceramic nanofeatures may then be synthesized. This technology is not entirely independent from hydrogen, bio-oil and other carbon containing gas capture processes currently under investigation from biomass sources by others in industry and academia. The highest obtained SSA value of ~933m^2g^{-1} did not exceed 1000m^2g^{-1}; less than 1/3 the highest reported result from the available literature[4]. The maximum SSA values could be considerably improved by refining material preparation for adsorption analysis as well as additional post-processing techniques.

The utilization of these material templates for further development of SiC structures necessitates retaining natural pore blocking substances within the biomass material which will develop carbon containing vapor upon pyrolysis. This will inherently reduce the measured SSA of the activated carbon templates produced in this work. Co-optimization of SSA values and nano-wire growth potential as well as tailored nanowire growth at specific locations within and upon the biomass carbon is under investigation by our group. Maximizing development of carbon containing gases during synthesis is a key to increase nanowire aspect ratio and yield. Manipulation of gas flow conditions, reduction of catalyst particle size, improved dispersion and reactant ratios during activation and further synthesis are significant factors for further optimization of nanowire growth.

The micro- and nano- porosity observed through SEM may not provide significant adsorption potential for hydrogen vapor; however, the hierarchy of pore structure may allow enhanced diffusion of hydrogen into the adsorbent bed nano- and sub-nano-structure observed by TEM. Micro- and sub-micro-pores may also provide optimal sites for the development of silicon carbide or other catalyzed nano-structures due to increased contact between material interfaces and localized environmental effects.

The current XRD analysis was not able to confirm the chemical composition of GICs or nano-flower features possibly due to relatively low concentration of the composite within the sample. Though we have not verified in this report the existence of potassium doped carbon materials, these compounds have been reported to increase the reactivity/ binding energy of carbon materials and provide interlayer spacing optimal for hydrogen adsorption[23].

Combining the considered material systems may realize a distinct and complimentary set of properties which are advantageous not only for adsorption of hydrogen, but for a number of other gas and particle applications. The utilization of agricultural-waste biomass may provide a more sustainable route to achieve tailored nano-structured materials with unique composite structure and synthesis potential.

ACKNOWLEDGEMENTS

The support of the National Science Foundation Division of Civil and Mechanical Systems and Manufacturing Innovations (NSF Grant no. CMMI-0758232) is greatly appreciated. Special thanks to Gordon and Lorna Bradbury of North Dakota State University Williston Research Extension Center (NDSU-WREC) for providing moral support and the precursor materials for this research. Support of Dr. McElfresh, Dr. Margaret Stern and Bob Melenson of SUN Microsystems San Diego is gratefully acknowledged. Another thank you to Dr. Stephen Barlow and Mariam Ghochani of the San Diego State University Electron Microscopy Facility for providing EM training as well as performing TEM characterization. Finally a thank you to my colleagues Evan Khaleghi, Wei Li and Gordon Brown of San Diego State University, Powder Technology Lab for their assistance and support during the course of this research.

REFERENCES

[1] X. Chen, O. Timpe, S.B.A. Hamid, R. Schlögl and D.S. Su, Direct Synthesis of Carbon Nanofibers on Modified Biomass-Derived Activated Carbon, *Carbon*, **47**, 340-47 (2008).
[2]M. Cox, A.A. Pichugin, E. I. El-Shafey, and Q. Appleton, Sorption of Precious Metals onto Chemically Prepared Carbon from Flax Shive, *Hydrometallurgy*, **78**, 137-144 (2005).
[3] F. Zhang, H. Ma, J. Chen, G.D. Li, Y. Zhang, and J.S. Chen, Preparation and gas storage of high surface area microporous carbon derived from biomass source cornstalks, *Bioresour. Technol.*, **99**, 4803-08 (2008).
[4] V. A. Yakovlev, P.M. Yeletsky, M. Yu. Levedev, D. Y. Ermakov, and V. N. Parmon, Preparation and investigation of nanostructured carbonaceous composites from the high-ash biomass, *Chem. Eng. J.*, **134**, 246-55 (2007).
[5] F. Cheng, J. Liang, J. Zhao, Z. Tao, and J. Chen, Biomass Waste-Derived Microporous Carbons with Controlled Texture and Enhanced Hydrogen Uptake, *Chem. Mater.*, **20**, 1889-95 (2008).
[6] X. Qin, X.P. Gao, H. Liu, H.T. Yuan, D. Y. Yan, W.L Gong and D.Y. Song, Electrochemical Hydrogen Storage of Multiwalled Carbon Nanotubes, *Electrochemical and Solid-State Letters*, **3**, 532-35 (2000).
[7] J.Y. Fan, X.L. Wu and P.K. Chu, Low-dimensional SiC nanostructures: Fabrication, Luminescence, and Electrical Properties, *Prog. Mater. Sci.*, **51**, 983-1031 (2006).
[8] G. Mpourmpakis, G. E. Froudakis, G. Lithoxoos and J. Samios, SiC Nanotubes: A Novel Material for Hydrogen Storage, *Nano Lett.*, **6**, 1581-83 (2006).
[9] J.J. Niu and J.N. Wang, Synthesis of Macroscopic SiC Nanowires at the Gram Level and their Electrochemical Activity with Pt Loadings, *Acta Mater.* **57**, 3084-90 (2009).
[10] R.A. He, Z.Y. Chu, X.D. Li and Y.M. Si, Synthesis and Hydrogen Storage Capacity of SiC Nanotube, *Key Eng. Mater.*, **368/372**, 647-49 (2008).
[11] S. Galvagno, S. Portofino, G. Casciaro, S. Casu, L. d`Aquino, M. Martino, A. Russo and G. Bezzi, Synthesis of beta Silicon Carbide Powders from Biomass Gasification Residue, *J. Mater. Sci.*, **42**, 6878-86 (2007).
[12] P. González, J. Serra, S. Liste, S. Chiussi, B. León, M. Pérez-Amor, J. Martinez-Fernández, A.R. de Arellano-López, and F.M. Varela-Feria, New biomorphic SiC ceramics coated with bioactive glass for biomedical applications, *Biomater.*, **24**, 4827-32 (2003).
[13] R. Xue and Z. Shen, Formation of Graphite-potassium Intercalation Compounds During Activation of MCMB with KOH, *Carbon*, **41**, 1862-64 (2003).
[14] R. Xue, H. Liu, P. Wang and Z. Shen, Formation of Nanocarbons During Activation of Mesocarbon Microbeads with Potassium Hydroxide, *Carbon,* **47**, 318-20 (2008).
[15] T. Kang, X. Liu, R.Q. Zhang, W.G. Hu, G. Cong, F. Zhao and Q. Zhu, InN Nanoflowers Grown by Metal Organic Chemical Vapor Deposition, *Appl. Phys. Lett.*, **89**, 071113 (2006).
[16] Z.W. Pan, Z.R. Dai, C. Ma and Z.L. Wang, Molten Gallium as a Catalyst for the Large-Scale Growth of Highly Aligned Silica Nanowires, *J. Am. Chem. Soc.*, **124**, 1817-22 (2001).
[17] G.W. Ho, A.S.W. Wong, A.T.S. Wee and M.E. Welland, Self-assembled Growth of Coaxial Crystalline Nanowires, *Nano Lett.,* **4**, 2023-26 (2004).

[18] C. Emmenegger, J.M. Bonard, P. Mauron, P. Sudan, A. Lepora, B. Groberty, A. Zuttel and L. Schlapbach, Synthesis of Carbon Nanotubes over Fe Catalyst on Aluminum and Suggested Growth Mechanism, *Carbon*, **42**, 539-47 (2003).
[19] S. Chakraborty, J. Chattopadhyay, W. Guo and W.E. Billups, Functionalization of Potassium Graphite, *Angew. Chem. Int. Ed.* **46**, 4486-88 (2007).
[20] E. Ziambaras, J. Kleis, E. Schröder and P. Hyldgaard, Potassium Intercalation in Graphite: A van der Waals Density-Functional Study, *Physical Review B*, **76**, 155425 (2007).
[21] G. Wei, W. Qin, K. Zheng, D. Zhang, J. Sun, J. Lin, R. Kim, G. Wang, P. Zhu and L. Wang, Synthesis and Properties of SiC/SiO_2 Nanochain Heterojunctions by Microwave Method, Cryst. *Growth Des.* **9**, 1431-35 (2009).
[22] J. Wei, K. Li, H. Li, D. Hou, Y. Zhang and C. Wang, Large-scale Synthesis and Photoluminescence Properties of Hexagonal-shaped SiC Nanowires, *J. Alloys Compd.* **462**, 271-74 (2008).
[23] R.J.M. Pellenq, F. Marinelli, J.D. Fuhr, F. Fernandez-Alonso and K. Refson, Strong Physisorption Site for H2 in K- and Li-doped Porous Carbons, *J. Chem. Phys.* **129**, 224701 (2008).

SYNTHESIS, STRUCTURAL AND MECHANICAL CHARACTERIZATION OF ARTIFICIAL NANOCOMPOSITES

Yong Sun, Zaiwang Huang, Xiaodong Li*
Department of Mechanical Engineering, University of South Carolina, 300 Main Street, Columbia, SC 29208, USA

*lixiao@engr.sc.edu

ABSTRACT

Nanoclay-reinforced polyacrylamide nanocomposite thin films were prepared through a facile electrophoretic deposition method. The re-aggregated nanoclay-platelets were identified to have a curled structure for releasing the strain against peeling off. Such a curvature is a possible source for realizing the interlocking mechanism in the nanocomposites. The localized deformation mechanism of the synthesized materials was investigated through micro-/nano-indentation and *in situ* observation of the deformation during tensile test with an atomic force microscope (AFM). Both indentation and tensile test results revealed that a localized crack diversion mechanism worked in the synthesized nanocomposite thin films, resembling its nature counterpart-nacre. Such a unique architecture provides a new approach to achieving localized mechanical reinforcing in nanocomposites. The deformation behavior and fracture mechanism are discussed with reference to lamellar structure, interfacial strength between the nanoclays and the polyacrylamide matrix, and nanoclay agglomeration.

INTRODUCTION

Nacre (mother-of-pearl) is one of the most amazing masterpieces created by Mother Nature, especially in the aspect of its mechanical property. It is a combination of ultra-high strength, and super-high toughness, which is a big surprise considering the ingredient: approximately 95% brittle inorganic aragonite (a mineral form of $CaCO_3$) and a few percent of soft inorganic biopolymer. The key to bypass the weakness of each component is to integrate platelet-like ceramic building blocks with a biopolymer layer between the blocks to render hybrid materials, also known as the brick-and-mortar arrangement [1-3]. Recently, it is further confirmed, through AFM observation and high-resolution transmission electron microscopy (HRTEM) study, that individual aragonite platelet is composed of numerous nanoparticles with an average size of 32-44 nm. Such self-assembled architecture exhibits a ductile rather than brittle characteristic and hence improves the overall robustness of the structure under mechanical loading [4-5], yielding a 2-fold increase in strength and a 1000-fold increase in toughness comparing to the constituted materials. The relationship between the exquisite structure and the fascinating mechanical behavior has been studied intensively, and substantial progress has been made on the understanding of the structure and mechanical reinforcing mechanisms involved in nacre [1-5].

All the studies reflect the keen desire of the community to comprehend, and finally, to mimic the secret recipe. Though as it always turns out, it remains a challenging goal to be an apprentice of Mother Nature. A number of different materials have been selected as reinforcing platelets to manufacture polymer-matrix artificial composites, such as graphite, SiC and montmorillonite (MMT) clays [6-7], etc. Various routines of fabrication have been proposed, including layer-by-layer deposition, electrostatic approach, evaporation induced self-assembly and electrophoretic deposition [6-9]. Significant increase has been achieved particularly at rather low platelet concentrations. On the other hand, the composites produced with a higher platelet concentration have shown notably deviated behavior from the theoretical predictions. Strong nanoclay-reinforced nanocomposites, with a 50 volume % (vol %) platelet concentration have been created through cross-linking the polymer matrix, exhibiting tensile strength up to 400 MPa. However, the ductility of natural nacre has not yet been achieved, which is presumably due to: 1) the difficulties in complete exfoliation of the MMT platelets and 2) poor bonding between the platelets and the interfaces [10].

Research reported by Kotov et al. [11, 12] partially revealed the link between the structural properties of the mimicking materials and the comparatively poor mechanical behavior. Cutting-edge surface-sensitive technologies have been applied to serve the purpose. Nevertheless, to our knowledge, *in situ* observations on the failure of the nanoclay-reinforced nanocomposites have never been carried out in micro-/nano-scale. The positive results from other materials, including laminated Si_3N_4 / BN composites [13, 14] demonstrate the possibility that such observations might shed light on why the mimic does not functionalize as effectively as its nature created original version and how people might improve it. AFM is highly applicable in studying the surface of the polymer-matrix nanocomposites due to its powerful resolution down to nanometer-scale in 3 dimensions. By integrating an *in-situ* mechanical testing stage with the AFM, this study presents an in-depth understanding of the small-scale deformation and fracture mechanisms of a MMT based artificial nacre synthesized through electrophoretic deposition [9]. The electrochemical method was chosen due to its facility, rapidness and its known capability of producing materials with complex geometry. The ordered brick-and-mortar structure was composed of MMT and polyacrylamide through electrophoretic deposition and polymerization by ultraviolet-radiation. The products exhibit similar properties when comparing to the nanoclay-reinforced agarose nanocomposites fabricated through a gelation method [15]. The nanocomposites were revealed to illustrate a crack diversion mechanism locally, which fails to carry on in a long range, presumably due to the lack of long-range order of the MMT platelets.

EXPERIMENTAL

Synthesis of Nanocomposites

The brick, which was a Na^+ type montmorillonite (Na-MMT), was produced by Southern Clay Products. The majority of clay particles are single-layer platelets before dispersion. According to the previous study, the platelets have an average length of 174.7 nm and average aspect ratio of 108.3, giving that the characteristic thickness is within the range of a couple of nanometers [14]. Acrylamide was chosen to be the intercalating agent and reacting monomer. The chemicals were applied without further purification. The whole procedure can be briefly described in three steps. Firstly, the Na-MMT was pre-treated in order to make the organophilic clay (O-MMT), which is a common step in such procedure to achieve better exfoliation. Secondly, electrophoretic deposition of O-MMT onto the electrode surface was carried out, with addition of sodium polymeta phosphate (SPP) as a surfactant. The electrodes in this case were two copper plates with commercial purity and their surfaces were pre-cleaned using hydrochloric acid and sodium hydroxide solutions. A uniform coating was achieved before the substrate was dipped into de-ionized water for cleaning and removal of the residue of the acrylamide monomers. Photopolymerization was the last step, during which the thin film would be put into a benzil/methanol solution and exposed under an ultraviolet (UV) light for 24 hours. The polymerized thin films would then be dried completely in a vacuum oven at 60 °C for another 24 hours before its peeling off from the substrate. The details of the procedure can be found elsewhere [9].

Characterization of Nanocomposites

An environmental scanning electron microscope (ESEM) FEI Quanta200 (FEI, Oregan, USA) and a transmission electron microscope (TEM) Hitachi H8000 (Hitachi, Japan) were used for microstructural characterization of the as-deposited nanocomposites. In order to prepare the TEM samples, two different methods were employed. In one method, the dried thin film was grinned into powder. Microtome was used in the other method to cut the thin film in order to observe the cross-section of the film.

A Vickers hardness tester Micromet-1 (Buehler, Illinois, USA) was applied to generate microcracks on both top and cross-sectional surfaces of the thin film in order to investigate the localized deformation mechanisms.

A custom-designed mechanical testing stage was employed to perform uniaxial tensile tests on the artificial nacre samples. The stage was equipped with a stepping motor which had a displacement resolution of 1.6 μm. It was integrated with an AFM Dimension 3100 (Veeco, New York, USA) for *in situ* observations on the nanoscale deformation of the thin films. Nanoindentation was also carried out on the nanocomposites using a Triboscope nanoindenter (Hysitron, Minnesota, USA) in conjunction with the AFM. Previous publications can be referred to for further details of the system [16].

RESULTS AND DISCUSSION

Structural Characterization

Figure 1a shows the optical image of both the as-deposited nanocomposite before going through the polymerization process (left) and the dried thin film which had been peeled off after polymerization (right). The homogeneous distribution of the MMT platelets is apparently reflected by the uniform color of the as-deposited film, which has a thickness of up to a few tens of micrometers. The layered structure is conveniently observed with the help of SEM on both top and cross-sectional surfaces of the thin films (Figs. 1b and 1c). While the MMT platelets were stacked in a well-defined order before the thin film was peeled off from the substrate, it was difficult for the weak bonding at the polymer-MMT interface to maintain such ordered structure after peeling off. In order to incorporate the released strain caused by the de-bonding of the thin film, certain amount of curvature in the architecture is essential and was found in the structure, as illustrated in Fig. 1d. To be more specific, the pre-exfoliated MMT platelets were re-aggregated into larger particles during the synthesis, and such particles were curled after the thin films were peeled off.

TEM images of the powder samples further confirm the existence of the MMT platelets inside the nanocomposites (Fig. 2a). As denoted by the arrows in the image, the boundaries of the platelets can be clearly identified. The contrast in the image is most likely due to the different thickness of the platelets, or the stacking of the platelets in some area. It should also be pointed out that the weak bonding has been an issue while making the cross-sectional microtome samples since the cutting force from the diamond saw will cause the de-bonding of the interface. Such influence can be noticed in Fig. 2b. Another characteristic shown in the image is that the some of the platelets are also curved. Based on the above observations, it can be seen that instead of the model proposed by Kotov [7], in which the platelets were considered to be flat, the aggregates of the platelets are actually arched in most situations, as shown in Fig. 3. This should be brought to attention due to the fact that such architecture might be able to provide an interlocking mechanism during the deformation and fracture of the nanocomposites, as claimed in nature nacre [2]. Such a mechanism has not been pointed out in the analysis of the nacre mimicking nanocomposites previously.

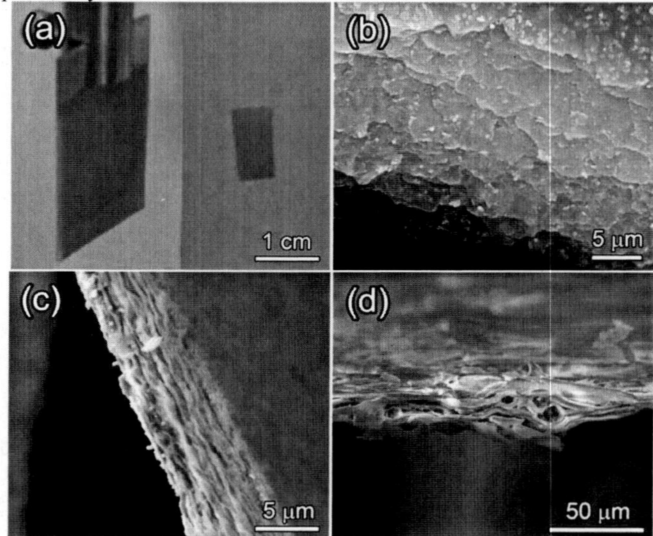

Figure 1. (a) Optical image of the as-deposited thin film before and after polymerization; (b) SEM image of the film top surface; (c)-(d) SEM images of the film cross-section before and after the thin film is peeled off from the electrode.

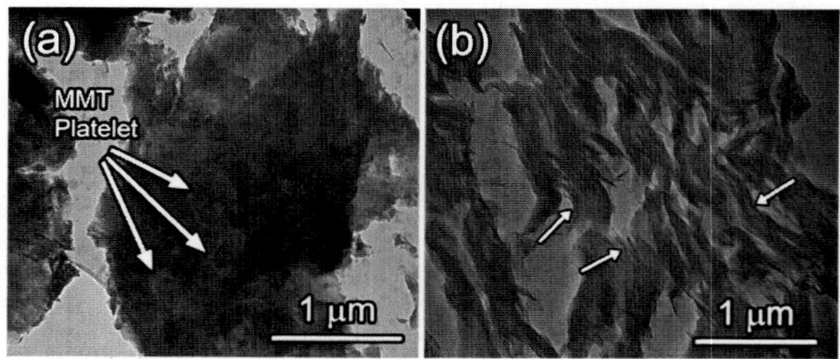

Figure 2. TEM image showing the representative morphology of (a) powder sample and (b) cross-sectional surface of the thin film.

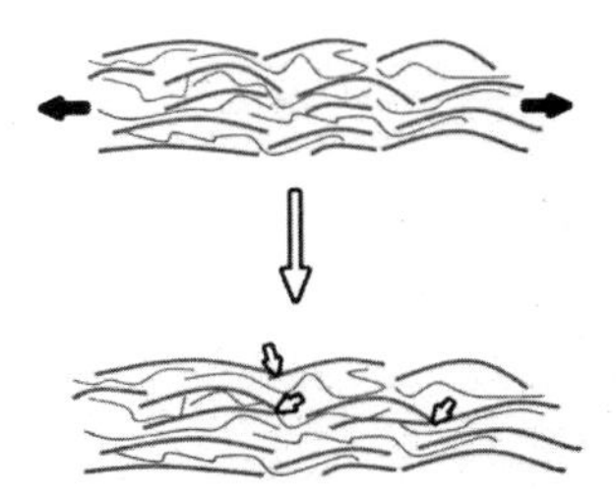

Figure 3. Curled aggregates of the MMT platelets leading to an interlocking mechanism (as shown by arrows) when the thin film is stretched.

Micro and Nano-Indentation

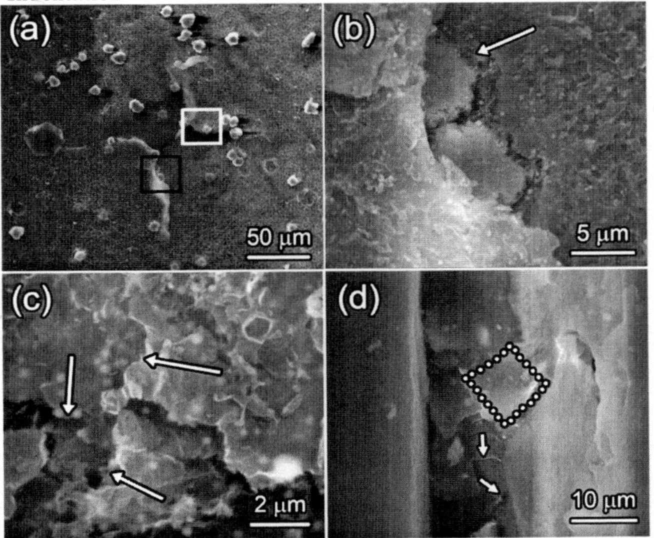

Figure 4. (a) A Vickers indentation mark on the top surface of the thin film. The localized deformation in the black and white boxes inside is represented by (b) and (c) respectively; (d) The indentation mark on the cross-section of the nanocomposites. Arrows denotes the secondary cracks developed.

Vickers indentation is always a versatile way to generate localized deformation and fracture, and to investigate the deformation and fracture mechanisms of materials. It has been proved to be effective in the research of nature nacre as well [17]. The microcracks produced by the indentation and their propagation paths provide the insights into the bonding between the polymer matrix and the reinforcement phase. A crack diversion mechanism was witnessed in the bottom and middle layers of the seashell of Pectinidae, which has a crossed lamellar structure. In both layers the cracks began within the indentation mark at its first-order lamellar interface, and then were deflected by the second and third-order lamellar structures, delicately enhancing the fracture toughness of the shell.

For the artificial nanocomposites, such crack diversion mechanism has been identified before, but not as effective as that in its nature counterpart, when observing the microcracks on the top surface of the thin film. This is due to the apparent lack of crossed-order structure. However, slight deviations of the primary cracks are obvious under SEM, as illustrated in Figs. 4b and 4c. One might also notice from Fig. 4b that the secondary cracks (which are noted by the arrow) propagates along what looks like the edge of the platelets. Fig. 4d, on the other hand, shows the microcracks on the cross-sectional surface. Contrary to the top surface, secondary cracks can be seen clearly, which are presumably formed continually along the platelet-polymer interfaces. As in the shell of Pectinidae, such secondary cracks play a crucial role by reducing the crack energy from the primary crack and constraining the shell damage by stopping failure occurring on a single interfacial plane. Such phenomenon is understandable since the stacking of the MMT platelets creates a lamellar structure on the cross-sectional area of the nanocomposites. From this point of view, it is also clear that the mechanical behavior of the nanoclay-reinforced nanocomposites is constrained by the shape of the nanoclay platelets, since such shape leads to the lack of lamellar structure on the plane perpendicular to the deposition direction.

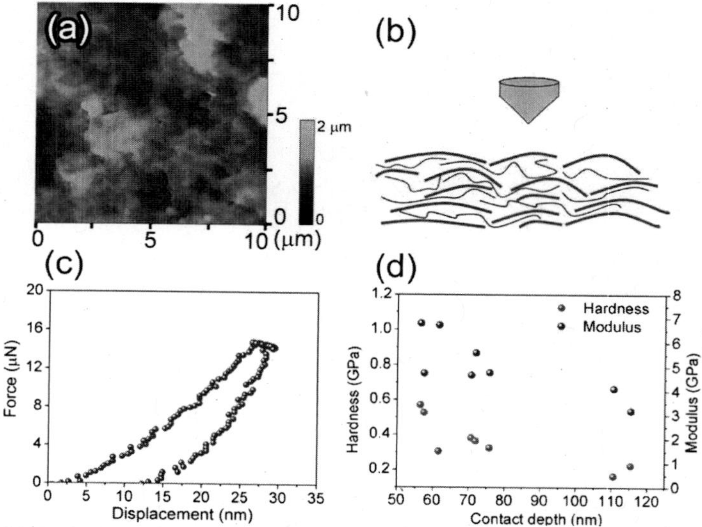

Figure 5. (a) Typical surface morphology acquired by AFM; (b) Schematic of nanoindentation on the thin film; (c) A representative force v.s. displacement curve of the nanoindentation and (d) Hardness and modulus values acquired with different indentation as a function of contact depth.

Nanoindentation was employed for probing the mechanical properties of the synthesized nanocomposites. The advantages of the technology include that long-range load transfer can be limited and the deformation zone is much smaller in nanoindentation comparing to traditional mechanical tests on a larger scale. Fig. 5a is a typical surface morphology of the nanocomposites acquired using AFM. Fig. 5d shows the hardness and elastic modulus of the nanocomposites measured by the nanoindentation tests, which are carried out in the direction parallel to the thin films. Various indentation forces were picked up so that the contact depths of the indentation marks fall in the range of approximately 60 – 110 nm, guaranteeing contacts between the tip and the MMT platelets whose typical thickness is only a couple of nanometers. The nanocomposites yield an average elastic modulus of 5.07 ± 1.24 GPa and a hardness of 0.36 ± 0.13 GPa, well comparable to our previous work done on the nanoclay-reinforced agarose nanocomposites [15]. It exhibits a great improvement compared with the polyacrylamide matrix. There is a clear trend of larger modulus / hardness for shallow contact depths. Considering the localized procedure of the indentation as illustrated in Fig. 5b, while the indenter tip encounters the MMT platelets, they might be bent down together with the tip, providing obstacles for the penetration. Or, the crack diversion mechanism can be in function as described in the previous section. Both can enhance the localized deformation behavior, which should be reflected in the results measured. However, as pointed out previously, the reinforcement effect in the nanocomposites cannot functionalize as well as that in nature nacre in the long range. The trend can serve as the evidence: as the indentation depth increases, the crack diversion mechanism fails to carry on when the tip encounters with more clay platelets, resulting in the decreased modulus and hardness.

In Situ Observation of Deformation and Fracture

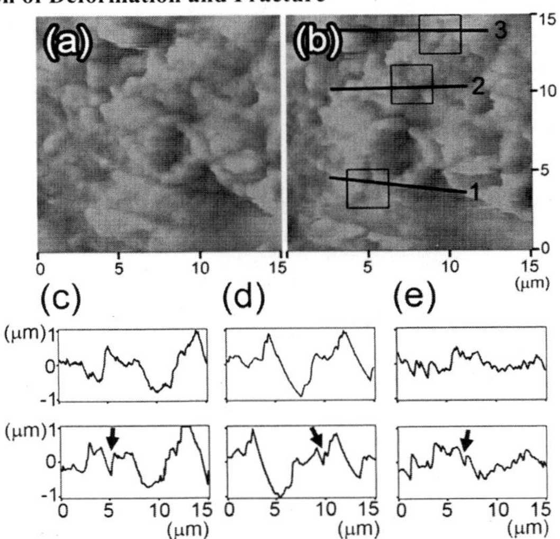

Figure 6. Surface morphology (a) before and (b) after the tensile test. The comparison of the height profiles on the three lines in (b) is represented by (c)-(e), respectively. The arrows in (c)-(e) reveal the positions of the microcracks.

By using the custom-designed tensile stage, localized deformation and fracture mechanisms can be investigated to further detail through analyzing the height profiles acquired with AFM. Figure 6 shows such an analysis in a 15 μm × 15 μm area. The sample was stretched horizontally for a strain of approximately 0.01 before the second image (Fig. 6b) was taken. The spherical site in the middle served as a source for strain concentration, which is also the reason why the specific location was picked for observation. Even without the help of the height profiles, initiation of microcracks around the spherical site can be seen in several directions. Three microcracks can be identified in the boxes marked in Fig. 6b, among which the microcrack along line 3 is believed to be a secondary crack generated by the deflection of the microcrack along line 2. This agrees with the results acquired from the micro-indentation tests, further confirming that the crack diversion mechanism stated.

From the height profiles, the procedures for the crack propagation can even be quantitatively analyzed. As illustrated in Fig. 6c, the first microcrack generated expanded to a width of ~ 300 nm and a depth of ~ 660 nm. The second microcrack (Fig. 6d), however, only spread out to a width of ~ 180 nm and a depth of ~ 360 nm. Several reasons can be the cause of such difference, and the generation of the secondary crack (Fig. 6e) is convinced to be one of them.

From the results from both micro-/nano-indentation and *in situ* observation under AFM, it is clear that the synthesized nanocomposites exhibit a localized crack diversion mechanism. However, it is also learned that the nacre-mimicking nanocomposites show a brittle behavior on fracture. The localized deformation mechanism fails to enhance the material on the long range, presumably due to the lack of ordered lamellar structure.

CONCLUSION

Nanoclay-reinforced polyacrylamide nanocomposite films were prepared through a facile electrophoretic deposition method. Through the structural characterization, it is seen that the nanoclay platelets were mainly homogeneously dispersed in the polyacrylamide matrix. Nanoindentation test shows that the nanocomposite thin films have an average elastic modulus of 5.07 ± 1.24 GPa and a hardness of 0.36 ± 0.13 GPa, which is improved greatly on the basis of the matrix. Micro-/nano-indentation together with *in situ* observation under tensile test reveal that the unique architecture in the thin films provides a localized crack diversion mechanism on both top and cross sectional surface, leading to the enhanced performance. Secondary cracks generated along the MMT platelets also improve the local mechanical properties of the material by reducing the crack energy. The trend for larger modulus / hardness with a shallow indentation depth and other observations give an insight into the poor transferring of the localized crack diversion mechanism. It is believed that the shape constraint of the clay platelets is the key to improve the overall performance of the nanocomposites. Lack of ordered lamellar structure results in the failure of the localized mechanism in a long range.

REFERENCES

[1] J.D. Currey, Mechanical Properties of Mother of Pearl in Tension, *Proc. R. Soc. London, Ser. B*, **196**, 443-63 (1977)

[2] A.P.Jackson, J.F. Vincent and R.M. Turner, The Mechanical Design of Nacre, *Proc. R. Soc. London, Ser. B*, **234**, 415-40 (1988)

[3] I.A. Aksay, M. Trau, S. Manne, I. Honma, N. Yao, L. Zhou, P. Fenter, P.M. Eisenberger and S.M. Gruner, Biomimetic Pathways for Assembling Inorganic Thin Film, *Science*, **273**, 892-8 (1996)

[4] X.D. Li, Z.H. Xu and R.Z. Wang, In Situ Observation of Nanograin Rotation and Deformation in Nacre, *Nano Lett.,* **6**, 2301-4 (2006)
[5] X.D. Li and Z.W. Huang, Unveiling the Formation Mechanism of Pseudo-Single-Crystal Aragonite Platelets in Nacre, *Phys. Rev. Lett.,* **102**, 075502-1-4 (2009)
[6] E.R. Kleinfeld and G.S. Ferguson, Stepwise Formation of Multilayered Nanostructural Films from Macromolecular Precursors, *Science,* **265**, 370-3 (1994)
[7] Z.Y. Tang, N.A. Kotov, S. Magonov and B. Ozturk, Nanostructured Artificial Naccre, *Nat. Mater.,* **2**, 413-8 (2003)
[8] M.H. Huang, B.S. Dunn, H. Soyez and J.I. Zink, In Situ Probing by Fluorescence Spectropy of the Formation of Continuous Highly Ordered Lamellar-Phase Mesostructured Thin Films, *Langmuir,* **14**, 7331-3 (1998)
[9] B. Long, C.A. Wang, W. Lin, Y. Huang and J.L. Sun, Polyacrylamide-clay Nacre-like Nanocomposites Prepared by Electrophoretic Deposition, *Compos. Sci. Tech.,* **67**, 2770-4 (2007)
[10] P.Podsiadlo, A.K. Kaushik and E.M. Arruda, Ultrastrong and Stiff Layered Polymer Nanocomposites, *Science,* **318**, 80-3 (2007)
[11] N.A. Kotov, T. Haraszti, L. Turi, G. Zavala, R.E. Geer, I. Dekany and J.H. Fendler, Mechanisms of and Defect Formation in the Self - Assembly of Polymeric Polycation - Montmorillonite Ultrathin Films, *J. Am. Chem. Soc.,* **119**, 6821-32 (1997)
[12] P. Podsiadlo, Z.Y. Tang, B.S. Shim and N.A. Kotov, Counterintuitive Effect of Molecular Strength and Role of Molecular Rigidity on Mechanical Properties of Layer-by-Layer Assembled Nanocomposites, *Nano Lett.,* **7**, 1224-31 (2007)
[13] Y. Sun, J. Liang, Z.H. Xu, G.F. Wang and X.D. Li, *In Situ* Observation of Small-Scale Deformation in a Lead-Free Solder Alloy, *J. Electron. Mater.,* **38**, 400-9 (2009)
[14] X.D Li, L.H. Zou, H. Ni, A.P. Reynolds, C.A. Wang and Y. Huang, Micro / Nanoscale Mechanical Characterization and *In Situ* Observation of Cracking of Laminated Si_3N_4 / BN Composites, *Mater Sci Engr C,* **28**, 1501-8 (2008)
[15] X.D. Li, H.S. Gao, W.A. Scrivens, D.L. Fei, V. Thakur, M.A. Sutton, A.P. Reynolds and M.L. Myrick, Structural and Mechanical Characterization of Nanoclay-Reinformced Agarose Nanocomposites, *Nanotechnology.,* **16**, 2020-9 (2005)
[16] X.D. Li, H.S. Gao, C.J. Murphy and L.F. Guo, Nanoindentation of Cu_2O Nanocubes, *Nano Lett.,* **4**, 1903-7 (2004)
[17] X.D. Li and P. Nardi, Micro/Nanomechanical Characterization of a Natural Nanocomposite Material - the Shell of Pectinidae, *Nanotechnology,* **15**, 211-7 (2004)

PROPERTIES OF FREEZE-CASTED COMPOSITES OF SILICA AND KAOLINITE

J. Walz
Department of Chemical Engineering
K. Lu
Department of Materials Science and Engineering
Virginia Polytechnic Institute and State University
Blacksburg, Virginia, USA

ABSTRACT

An experimental study was conducted to investigate the effects of added micron-sized kaolinite clay particles on the properties of porous, freeze-dried composites of clay and silica nanoparticles. Adding salt (sodium chloride) to aqueous mixtures of clay and silica produces a sol-to-gel transition in which the clay particles are dispersed. The presence of the clay particles increases significantly the rate of the gellation. SEM images of the dried composites show a porous material with pore sizes that decrease with increasing rate of freezing. Surface area measurements show that the freezing rate may alter the pore structure of the sample, with faster freezing producing smaller and more numerous pores.

INTRODUCTION

Clays are unique particles with plate-like geometries. The particle diameters are on the order of microns and thicknesses in the nanometer range. Kaolinite is one of the most common types of clays and is formed of several hundred layers of alumina- and silica-like materials which are covalently and ionically bonded. Natural synthesis leads to ionic substitution in the crystal structure, and thus mined clays are usually crystallographically imperfect and quite polydisperse in nature. Because of the alternating silica- and alumina-like platelet structure, kaolinite has one face alumina-like and the other silica-like. The relationship between surface charge and solution pH is different for each of these materials, and experiments have found the platelet faces to be negatively-charged and the edges positively-charged.[1]

Recently, we discovered a unique sol-to-gel transition that was produced in aqueous kaolinite suspensions upon the addition of sufficient amounts of electrolyte and silica nanoparticles.[2,3] SEM micrographs showed that the clay platelets in the gel were arranged in a 'honeycomb' structure consisting of edge-to-face contact of the platelets and pore sizes of 1 – 2 μm. In addition, the gels displayed a significant yield stress and could repeatedly and reproducibly re-form after breakage by shear with no significant change in properties.

From a different perspective, gellation of a dispersion of silica nanoparticles can be produced by the addition of an appropriate amount of sodium chloride. In aqueous solutions above pH of approximately two, silica nanoparticles are stabilized against aggregation by negative charges on the surfaces. The addition of an electrolyte, such as sodium chloride, screens the electrostatic repulsion resulting from this surface charge, allowing the nanoparticles to aggregate into a network of interconnected particles held together by van der Waals attractive forces.[4] When relatively large kaolinite particles are present in the initial dispersion, the particles become trapped within the gelling network.

While significant work on the properties of freeze-casted materials has been performed, the use of added clay particles to alter the properties of the resulting composite has not been studied. In previous work, we investigated the effects of various sized silica nanoparticles on the rheological properties of the silica/kaolinite gels, and the pore structure of the freeze-dried composite.[2,3] In the current paper, our objective is study the effects of the specific freezing process used, as well as investigate the effects of the clay on the material properties of the freeze-dried composites.

Historically, clays have found use in one of two applications. They are the primary component of ceramics, where they impart insulative properties and can be heated to form refractory materials. As an additive, they are used for their unique anisotropic particle shape, which modifies rheological properties of materials such as paints, paper coatings, and polymers. One potential application of the silica-kaolinite gels is impact protection, in the form of insert packaging of helmets. The unique microstructure of the silica-clay systems at solid state in this study also offers numerous potential applications, such as catalyst supports, insulating materials, and filters. Composites made from such compositions will be lightweight and chemically inert, thus possessing potentially useful properties for a wide range of technological applications, including catalyst supports, filters, membranes, and heat insulating materials.

EXPERIMENTAL PROCEDURE

Materials

In this study, kaolinite clay (Hydrite Flat-D, Imerys Performance Materials, South Carolina), Ludox TMA (Sigma Aldrich), sodium chloride (AR grade, Mallinckrodt) and distilled water were used to produce the composites. Ludox is a registered trademark of E.I. DuPont de Nemours & Co., Inc., and TMA describes the product as silica nanoparticles with a mean diameter of 22 nm. The material is supplied as a 34 wt% suspension in de-ionized water (pH 7.3). The kaolinite is supplied as a dry solid, has a manufacture-listed specific surface area of 7 m^2/g, and is polydisperse with a mean plate diameter of 5 μm and a measured particle thickness of roughly 100-200 nm. The density of the clay was determined by displacement in water as 2.56 g/cm^3.

Prior to use, the kaolinite was washed in a 1 M NaCl solution at pH 3 according to the method of Schofield and Samson.[5] This washing served essentially three purposes. First, it created a more uniform surface chemistry on the platelets. Second, it removed any dispersants that remained in the stock solution after preparation by the manufacturer. And third, it ensured that the primary counterions in the solution were sodium and chloride. After centrifugation at 6000 RPM, the supernatant was discarded and the solid was re-dispersed in a fresh solution. This was repeated seven times and then the process was repeated with de-ionized water. The pH was controlled using 1M HCl (Mallinckrodt).

After the washing steps in de-ionized water, the clay was dried in an oven overnight at 100°C to remove excess water and then stored in a closed container until use. Conductivity measurements on suspensions of washed clay in de-ionized water at 14 vol% showed an increase in NaCl concentration, at approximately 3 mM. This value is well below the 500 mM NaCl used to induce gellation, meaning that the amounts of sodium and chloride ions present in the gelling solutions were well controlled. Both the sodium chloride and Ludox TMA were used as obtained from the manufacturer.

Sample Preparation

Suspensions were prepared in de-ionized water with varying volume fractions of clay, silica and concentration of salt. All samples were prepared by adding the Ludox suspension to de-ionized water, followed by the addition of kaolinite. The suspension was homogenized using a vortexer for 5 minutes and then an appropriate volume of a 5 M NaCl solution was added to bring the salt concentration and total volume to the desired level. The suspension was then agitated again for 5 minutes. The notation describing samples in this paper are described by the percent volume clay, followed by percent volume silica, and then by concentration of salt. For example a 14% by volume clay, 8% by volume silica, and 0.5 M NaCl sample is written as 14/8/0.5. The remaining volume percentage is water.

Immediately after the last agitation step, the suspensions were poured into molds and allowed to gel in a closed container for 4-8 hrs depending on the concentration of clay. Samples containing no clay were gelled for 8 hrs. The samples were rapidly submerged in either liquid nitrogen or liquid propane (in a bath of liquid nitrogen) and subsequently transferred to a bell jar while submerged in liquid nitrogen (frozen samples were transferred to a fresh container of liquid nitrogen to ensure melting did not occur prior to vacuum exposure). The samples were then exposed to vacuum overnight to sublimate the water. Additionally, some samples were frozen using a freeze drier (Labconco Stoppering Tray Dryer, Labconco, Kansas City, MO) at -35°C. The freeze-drier has an incorporated vacuum system, so no sample transfer occurred between the freezing and sublimation steps.

Characterization

Scanning Electron Microscope images were obtained using an FEI Quanta 600 FEG microscope (FEI Company, Hillsboro, Oregon). Freeze-dried samples were fractured and images were taken from the fractured face to observe the bulk structure. The exposed face was coated in 15 nm of gold prior to imaging. BET surface area analysis was conducted on small fractured samples using an Autosorb-1 (Quantachrome Instruments, Boynton Beach, FL).[6,7] Care was taken not to damage the bulk integrity of the sample so that the porous structure of the samples and the resulting surface area were not altered.

RESULTS AND DISCUSSION

Gellation of Nanoparticle/Kaolinite Suspensions

A general phase diagram based on the volume fractions of Ludox TMA, with and without kaolinite, as a function of the concentration of NaCl, is presented in Fig. 1. The lower left quadrant represents composites that did not gel, up to and including the values of the data point. The upper right quadrant signifies composites that resulted in a stable gel. The addition of 7 vol% and 14 vol% kaolinite did not significantly affect the boundary between rigid gels and liquid composites, suggesting that the gellation is driven primarily by the silica nanoparticles.

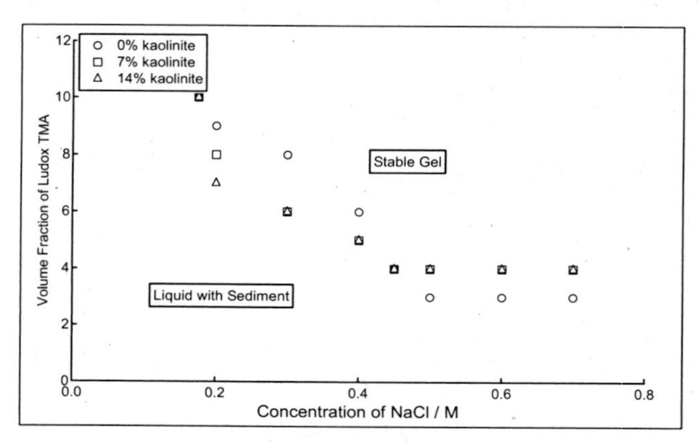

Figure 1. General phase diagram for the gellation of silica nanoparticles as a function of the concentration of NaCl, with and without added kaolinite. The addition of kaolinite has a minimal effect on the gelation of the Ludox as a function of salt.

The addition of clay does, however, affect the sol-gel transition rate. This casual observation was quantitatively confirmed using a rheometer to measure the storage modulus (G') as a function of time (Fig. 2). The results shown here were measured at 25°C between two parallel plates at a separation of 1 mm, using dynamic oscillation at a strain rate of 0.2% (within linear visco-elastic region) and an oscillating frequency of 1 Hz. The Y-axis in Fig. 2 is the ratio of the measured storage modulus to the modulus measured at the end of eight hours, referred to as G'_{max}. It should be noted that this eight-hour modulus is not the absolute maximum because the sample continues to slowly strengthen some beyond this time; it was simply the last measured value at the end of the experiment. After one hour, the nanoparticle/kaolinite suspension had achieved ~ 28% of its maximum gel strength (G'_{max} = 4000 Pa) and was strong enough to resist flow if the container was tipped upside down, whereas the silica-only sample had only achieved ~ 3% of its maximum gel strength (G'_{max} =530 Pa) and still flowed if the container was upended. The addition of the kaolinite increases the storage modulus of the gel over the time frame studied. Whether this trend would continue over longer periods of time is not known. In addition, it should be stated that the exact cause of this increase is currently not understood and is still being investigated.

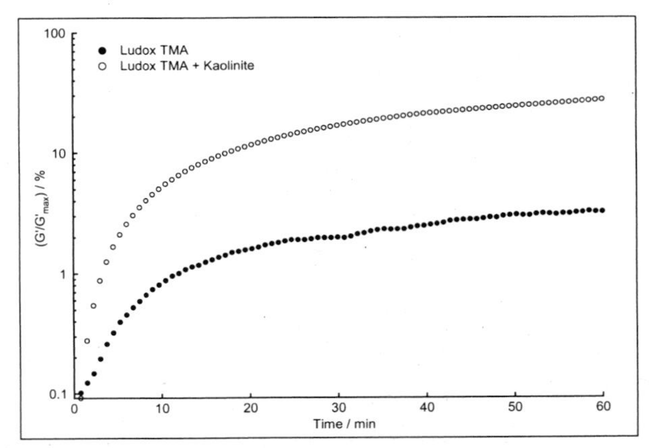

Figure 2. Percent storage modulus of 0/8/0.5 and 14/8/0.5 suspensions measured as a function of time. G'_{max} for the Ludox TMA and Ludox TMA+kaolinite suspensions at the end of 8 hours was 530 and 4000 Pascals respectively. After one hour of gelation, the nanoparticle-clay mixture behaved as a solid and would not pour from the container, whereas the nanoparticle-only suspension would still flow.

Below 7 vol% kaolinite, the settling rate of the clay was faster than the rate of gellation, which leads to a phase separated gel. This is likely because that the addition of kaolinite above 7 vol% increases the rate of gellation by increasing the rate of interconnected-network growth. The gellation rate of a suspension of small particles is dependent upon the time it takes for individual particles to collide, followed by collision with other monomers, or dimers, or trimers (assuming that each collision is irreversible), which will continue until a volume-spanning network is achieved. With each successive increase in aggregate size, the probability that the Brownian rotation of this aggregate will lead to another contact, increases. For example, the rotation of a spherical object will not decrease the distance between itself and its next nearest neighbor; however, the rotation of an irregular shaped aggregate, such as a disk-shaped clay platelet, can decrease the distance between neighbors by rotation, increasing the rate of collisions and network growth. The addition of clay platelets likely increases the rate of interconnected network growth by supplying a large, non-spherical volume-spanning object, whose rotation has the ability to decrease the time it takes for collisions to occur. However if the volume fraction of clay is too low, then settling of the clay occurs faster than a volume-spanning rigid network can be attained; resulting in a phase separated gel of silica resting on top of aggregated clay.

Effect of Freezing Rate on Pore Structure and Surface Area

Figs. 3A, B and C, demonstrate the effect of freezing rate on the pore structure of a 9/8/0.5 composite sample. Each sample volume was approximately 2 cm^3. Sample A (Fig. 3A), was slowly brought to -35°C using the Labconco freeze-drier and then sublimated under vacuum. Samples B and C were rapidly frozen by submersion in either liquid nitrogen (Fig. 3B) or liquid propane that was itself in a bath of liquid nitrogen (Fig. 3C) and then sublimated. The efficiency of liquid nitrogen to freeze an object is diminished by rapid boiling upon introduction of a warm object to the liquid, also known as the Leidenfrost effect.[8] This creates a vapor barrier between the object being frozen and the cold liquid, effectively insulating the object. This effect is diminished in liquid propane because its boiling point (-42°C) is much higher than that of liquid nitrogen (-196°C). Samples dipped in liquid propane at -196 °C therefore freeze at a faster rate. The decreasing pore size demonstrates that increasing the rate of freezing results in a larger number of smaller ice crystals, as evidenced by smaller and more numerous pores. As the rate of freezing increases, the pore size decreases. Comparing samples 4B and 4D demonstrates that increasing the volume fraction of clay decreases the pore size. This is further demonstrated in sample 4E, where no clay is present.

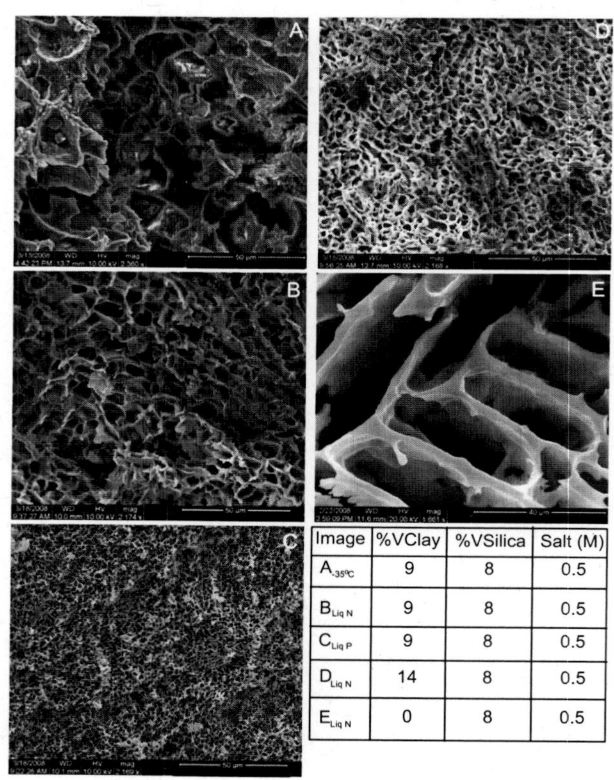

Image	%VClay	%VSilica	Salt (M)
$A_{-35°C}$	9	8	0.5
$B_{Liq\ N}$	9	8	0.5
$C_{Liq\ P}$	9	8	0.5
$D_{Liq\ N}$	14	8	0.5
$E_{Liq\ N}$	0	8	0.5

Figure 3. SEM images of freeze dried composites prepared with varying compositions and freezing methods (see table below).

Figs. 3D and B show that freezing samples in liquid nitrogen and decreasing the solid clay content from 14 vol% to 9 vol% at constant volume fraction of silica (8%), results in a slight increase in pore size. At 0 vol% clay and 8 vol% silica (Fig. 3E) the size of the pores has increased substantially, and we also observe long range directional freezing over hundreds of microns. We suspect that measurements made with clay concentrations between 0 vol% and 9 vol% would show an inverse trend between clay concentration and pore size, possibly due to the relatively large platelets acting as barriers limiting the size and propagation of growing ice crystals. However, as mentioned above, actual settling of the platelets prior to the onset of gellation precluded this experiment.

The surface area of 9/8/0.5 and 14/8/0.5 composites, frozen in liquid nitrogen, was measured by adsorption of nitrogen gas and quantified through BET analysis. Fig. 4 shows an example adsorption and desorption branch for nitrogen as a function of the relative pressure (the inset is the BET plot) adsorbed on a 9/8/0.5, freeze-dried composite. The hysteresis loop confirms a porous material and the specific surface area was calculated to be 64.5 m^2/g. Over multiple runs the specific surface area was 62 ± 4 m^2/g. The adsorption-desorption branch for the 14 vol% clay is similar in shape (not plotted) with a measured specific surface area of 52 ± 2 m^2/g. These values are consistent with the known properties of the Ludox TMA nanoparticles (specific surface area of 140 m^2/g, density of 2.4 g/cm^3) and kaolinite (specific surface area of 7 m^2/g, density of 2.56 g/cm^3) with the assumption that the total surface area is conserved upon mixing.

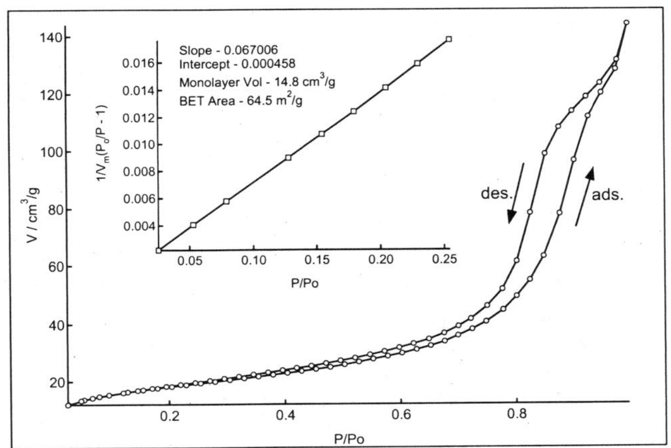

Figure 4. Adsorption and desorption branch of nitrogen onto a 9/8/0.5 composite. The BET surface area plot is in the inset. The monolayer volume and specific surface area are 14.8 cm^3/g and 64.5 m^2/g respectively. Hysteresis between the adsorption and desorption branch confirms the sample is porous.

CONCLUSION

In this work, Ludox TMA SiO_2 nanoparticles have been used to suspend kaolinite clay in a gel state and freeze-drying has been used to produce a highly porous solid. The presence of the clay particles increases significantly the rate of the gellation. SEM images of the dried composites show a porous material with pore sizes that decrease with increasing rate of freezing. The pore structure of the sample depends on both the rate of freezing and volume fraction of suspended kaolinite. Surface area measurements show that the freeze rate may alter the pore structure of the sample, with faster freezing producing smaller and more numerous pores.

ACKNOWLEDGMENT

The authors acknowledge the financial support from National Science Foundation under grant No. CBET-0827246.

REFERENCES

[1]H. Van Olphen, *An Introduction to Clay Colloid Chemistry*, 2nd ed.; Wiley-Interscience: New York, 1977.

[2]J. C. Baird and J. Y. Walz, The effects of added nanoparticles on aqueous kaolinite suspensions I. Structural effects, *J. Colloid Interf. Sci.*, **297**, 161-169 (2006).

[3]J. C. Baird and J. Y. Walz, The effects of added nanoparticles on aqueous kaolinite suspensions II. Rheological effects, *J. Colloid Interf. Sci.*, **306**, 411-420 (2007).

[4]L. L. Hench and J. K. West, The sol-gel process, *Chem. Rev.*, **90,** 33-72 (1990).

[5]R. K. Schofield and H. R. Samson, Flocculation of kaolinite due to the attraction of oppositely charged crystal faces, *Discuss. Faraday Soc.*, **18,** 135 (1954).

[6]S. Brunauer, P. H. Emmett, and E. Teller, Adsorption of gases in multimolecular layers, *J. Am. Ceram. Soc.*, **60,** 309 (1938).

[7]S. Lowell and J. E. Shields, *Powder Surface Area and Porosity*, in 3rd ed.: Chapman & Hall Ltd. (1991).

[8]S. Deville, Freeze-casting of porous ceramics: A review of current achievements and issues, *Adv. Eng. Mater.*, **10**, 155-169 (2008).

CONTROLLED PROCESSING OF BULK ASSEMBLING OF NANO-PARTICLES OF TITANIA

M. Jitianu[1], J.K. Ko[1], S. Miller[1], C. Rohn[2], R. A. Haber[1]
[1]Rutgers, The State University of New Jersey, 607 Taylor Rd, Piscataway, NJ 08854, USA

[2]Malvern Instruments Inc., 117 Flanders Road, Westborough, MA 01581-1042, USA

ABSTRACT

The use of nanomaterials to form bulk shapes has gained significant attention due to their unique properties with regards to conventional materials. The present investigation involves an aggregated nano-TiO_2 synthesized through a sulfate process. This route is providing opportunities to produce powders of varying starting properties for further use as catalysts. Generally, the manufacturing of a catalyst involves blending of the nano-TiO_2 with a binder and lubricant to form a paste prior to extrusion in desired shapes. The extrudability of a paste is strongly dependent on the physical and chemical characteristics of the bulk paste, thus equally on the source material and its additives. The effects of different organic additives along with the aggregation characteristics of the nano-TiO_2 employed on the extrusion of a paste and on the morphological characteristics of the final extrudate have been investigated and jointly correlated. Capillary rheometry has been employed. This technique is very suitable for describing and predicting the flow of particulate pastes during extrusion. The key role in controlling aggregation under shear was found to be held by the organic additives, specifically their different interaction with the TiO_2 aggregates.

INTRODUCTION

Titanium dioxide (TiO_2), particularly anatase, is broadly used in catalysis as filters, adsorbents and catalyst supports[1] and in pigments and composites industry, as well[2]. Moreover, nanostructured titania has received increasing attention because of photocatalytic and photovoltaic application[3]. Agglomeration characteristics of TiO_2 particles are of great importance, when preparing catalyst pellets to facilitate reactant/product flow[4].

The extrusion process for forming ceramics is widely employed in environmental applications and specifically in automotive and stationary NO_x emission control[5]. Extrusion has long been used successfully for the preparation of TiO_2-based catalysts, as well. Ceramic bodies of extremely complicated shapes are extruded, e.g. honeycomb structures with a cell density of up to 400 cells/in^2 [6].

The extrusion of ceramics requires addition of organic additives (binders) to a powder in order to produce a paste. To control the production, it is interesting to model behavior of the paste under conditions that are close to the actual extrusion conditions[7]. Each of the production steps in the extrusion process influences the quality of the final material. Thus, it is very important to know correlations between process parameters and the final properties of the catalysts[8]. Several aspects of the extrusion process have been addressed, including the relation between rheology and extrudability for alumina-based pastes, the dependence of TiO_2-based catalysts on the composition of the corresponding paste, mathematical modeling of the extrusion of ceramic pastes[9-11].

The quality of the final extrudate is controlled essentially by the chemical formulation and the rheological properties of the paste components, as well as the geometrical characteristics of the employed die. All these parameters have to be optimized in order that flow defects during extrusion are eliminated. For successful extrusion of ceramics, adequate plasticity must be provided by the addition of organic binders. Cellulose is one of the world's most abundant polymers, it is nontoxic and biodegradable, whereby its derivatives are widely employed. Studies have proven that cellulose-derivatives are beneficial as organic binders in the extrusion process[12,13]. Inorganic binders have the disadvantage that they cause serious contamination in the final product. Organic binders (polymers) are combusted in the pellet melting process, thus strongly reducing the amount of contamination. This application is a consequence of their adsorption behavior at a solid-liquid interface[14] . Most of the cellulose polymers are water-soluble and a very important property of their aqueous solutions is the water-holding capacity they exhibit[15]. Generally, methylcellulose (MC) and partially-substituted methylcellulose (HPMC), such as hydroxypropyl methylcellulose, are widely employed as organic additives in extrusion and injection molding of traditional ceramics[16,17]. Their solutions are generally pseudoplastic. This property is generally desirable to the extrusion process, where a low viscosity is required for the reduction of the applied pressure, while at low shear rates a high viscosity is required to prevent deformation of extrudates during drying. An important representative of the cellulose derivatives is carboxymethyl cellulose (CMC). CMC is used as an additive in papermaking, in pharmaceuticals and cosmetics, and in food products[18-20]. Especially CMC has been successfully applied as an alternative for inorganic binders in the pelletization process[21]. CMC has a low solubility in water, unless transformed into its corresponding sodium salt, Na-CMC. Regardless of its limited solubility, the applications are numerous and CMC is one of the world's largest produced polymers.

The aim of this present study is to correlate the rheological characteristics of titania-based pastes prepared with water soluble low-viscosity HPMC and low water soluble (almost insoluble) CMC, respectively, with the flow behavior of the ceramic paste, in order to foresee the extrudability of aggregated nano-TiO_2. Considering that extrusion can be envisaged as an ordinary flow process, it is useful to use rheological methods to investigate the formability behavior of ceramic pastes[22]. Capillary rheometry is generally employed for this purpose in view of its correspondence to the extrusion machines, since it is possible to analyze a range of shear rate values (1-10,000 s^{-1}) that compares well to those estimated during the extrusion process. The effects of the paste composition (binder employed) along with the aggregation characteristics of the nano-TiO_2 used on the rheology of the paste and on the morphological characteristics of the final extrudate have been investigated and mutually correlated.

EXPERIMENTAL

TiO_2 used in this work is commercially available and synthesized from ilmenite ore via sulfate process and consists of anatase with 1.8% residual sulfate wt%, expressed as SO_3 %.

Two cellulose-derivative binders have been employed, as follows: water low-soluble Carboxymethyl ether cellulose (CM-Cellulose, Sigma-Aldrich), further named CMC and water soluble hydroxypropyl methylcellulose (Dow Chemical), containing 5% hydroxypropyl wt%, further named HPMC.

A Bohlin Gemini rotational rheometer (Malvern Instruments Ltd. UK) was employed to carry out rheological tests on the cellulose-derivative binders' solutions. The cup and bob geometry was used for the viscosity tests, which have been conducted at T=25°C, temperature accurately controlled by a Peltier Couette system. The cup and bob consists of two concentric cylinders (25mm diameter, calibrated for 10mL solution), the outer cup and the inner bob with the test fluid in the annular gap. A solvent trap was used to avoid sample drying out. 10 mL solution of 2% binder wt% concentration in deionized water has been employed. The CMC suspension in water was prepared by adding CMC into water and stirring it continuously prior to be transferred to the rheometer. The HPMC solution was prepared as follows: water was brought to 70°C and the suitable amount of HPMC was then added under vigorous stirring. After HPMC was dissolved, the solution was cooled to room temperature and then kept overnight at 5°C. A transparent gel was effectively obtained using this procedure. The gel was brought back to 25°C prior to be used in paste preparation. Often called the "hot/cold" technique, this method takes advantage of the insolubility of HPMC in hot water, but an increased quality of dispersion and wetting of particles can be achieved at 70°C. After a thorough dispersion at this temperature, the suspension is cooled towards room temperature. Once the dispersion reaches the temperature at which HPMC becomes water soluble, the powder begins to hydrate and viscosity increases. Complete hydration, improved clarity and accurate control of viscosity of HPMC solutions depend on adequate cooling, generally carried out at 5°C[15].

The viscosity of CMC suspension and HPMC gel was measured as a function of shear rate. The shear rate was uniformly increased and then decreased over the range of 2 to 100 s^{-1}, with a total time period of 3 minutes for both the increasing and the decreasing shear rate sweeps.

Particle size distribution of titania starting powder and titania extrudates along with the CMC binder was carried out on a Mastersizer 2000 (Malvern Instruments Ltd. UK), in suspension (Hydro 2000S attachment). Conversely, the particle size distribution of the HPMC binder was carried out only using dry powder (Scirroco 2000 attachment), because of its high solubility in water. Measurements of particle size distribution of CMC binder were carried out using dry powder as well (Scirroco 2000 attachment), for comparison.

A Rosand 2002 capillary rheometer (Malvern Instruments Ltd. UK) was employed to measure the flow properties of the titania pastes. All measurements were conducted at T=25°C. Pastes have been prepared by mixing of all components for 30 minutes in a Haake Rheomix 600 Batch Intensive Mixer with Rheocord 9000 Control System. The binders have been pre-dispersed (CMC) or dissolved in water as described above (HPMC) and then added to the mixer along with the powder. Two pastes have been investigated with the following compositions: (1) 68% TiO_2 wt%, 4% CMC wt%, and 28% water wt% and (2) 68% TiO_2 wt%, 2% HPMC wt% and 30% water wt%. The proportions of the binders in each paste have been chosen such as the maximum torque value of the mixer unit to be very close for the two pastes.

Capillary flow tests were carried out at two constant apparent shear rates, 30 and 300 s^{-1}, respectively. Two flat-entry cylindrical dies of 1mm diameter with different L/D ratios 16 and 32, respectively have been used in the capillary testing, pastes being compressed through a barrel of 15mm diameter. The resulting pressure drop (total extrusion pressure) was then recorded for each speed and dies employed. The data have been then used to calculate extrusion parameters using Benbow and Bridgwater methodology[23]. Rabinowitsch correction for non-Newtonian fluid has been applied to calculate the true shear rate[7], accurate values for the pressure drop being further obtained and used in the calculation of the extrusion parameters.

Surface area of the TiO_2 powder and extrudates has been determined by $He\text{-}N_2$ adsorption-desorption using a Micromeritics Gemini V instrument.

SEM secondary electron images of the starting titania powder and the corresponding extrudates have been recorded on a Zeiss Sigma FESEM at 10kV, different magnifications being used, only 250,000x being shown further in the paper. Extrudates obtained using the die with L/D=16 at 10 mm extrusion speed have been analyzed by FESEM. This way, the aggregate structure of titania after applying high shear could be observed and compared to the pristine powder aggregate structure.

RESULTS AND DISCUSSION

Rheology of solutions of cellulose-derivatives plays an important role in many practical applications, where the modification of flow behavior is essential. A non-Newtonian fluid is one whose viscosity is independent on shear rate, while in practice, many systems exhibit non-Newtonian flow behavior, where viscosity may decrease (pseudoplastic) or increase (dilatant) with increasing rate of shear.

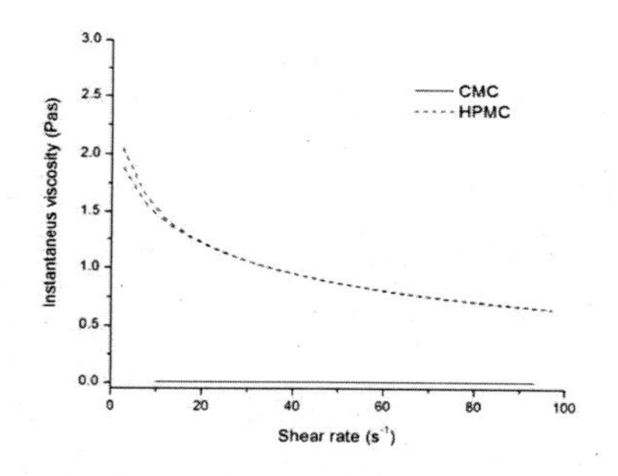

Figure 1. Viscosity as a function of shear rate for 2% binder (CMC and HPMC) in water.

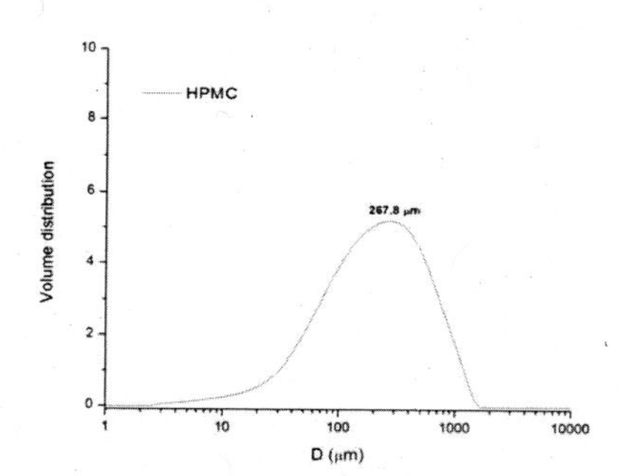

Figure 2. Particle size distribution of solid HPMC.

As seen in Figure 1, which illustrates the viscosity plots as a function of shear rate for both binders, the gel solution of 2% HPMC in water exhibits pseudoplastic or shear thinning behavior, leading to a lower viscosity as shear rate increased. The solution of CMC acts as a Newtonian fluid and it is constant throughout the whole shear range tested. As a consequence of the low solubility of CMC in water, the suspension of CMC exhibits a very low viscosity, 0.006 Pas. However, for water soluble methyl cellulose and derivatives, rheology of aqueous solutions is affected by many characteristics, such as polymer molecular weight, and its concentration[15].

Generally, pseudoplasticity increases with molecular weight or concentration. A general accepted nomenclature for cellulose compounds is based on the amount of substituent groups on the anhydroglucose units of cellulose[15]. This can be designated by weight percent or by the average number of substituent groups attached to the ring, a concept known to cellulose chemists as "degree of substitution" (D.S). If all three available positions on each unit are substituted, the D.S. is designated as 3; if an average of two on each ring is reacted, the D.S. is designated as 2, and so on. The number of substituent groups on the ring determines the properties of the various such compounds. A D.S. comprised between 1.64 and 1.92 yields maximum solubility in water, while a lower D.S. gives compounds with lower water solubility. Conversely, higher degrees of substitution produce polymers that are soluble only in organic solvents. Various degrees of substitution affect the molecular mass of the cellulose polymers, hence the rheology of their aqueous solutions. The higher the molecular weight, the more viscous the solutions will be. The HPMC employed in this paper has a D.S. equal to 1.8, a number average molecular weight of ~67,000. By extrapolation of the viscosity data in Figure 1, the zero shear viscosity of HPMC was found as 1.98 Pas. This value lays this polymer in the category of soluble cellulose polymers with low viscosity. Conversely, the D.S. of CMCs varies with the preparation procedure, but it is generally found between 0.6-0.95, corresponding to low water solubility.

Particle size distribution of the as-received polymers is presented in Figures 2 and 3, while particle size distribution of a water suspension of CMC is displayed in Figure 4.

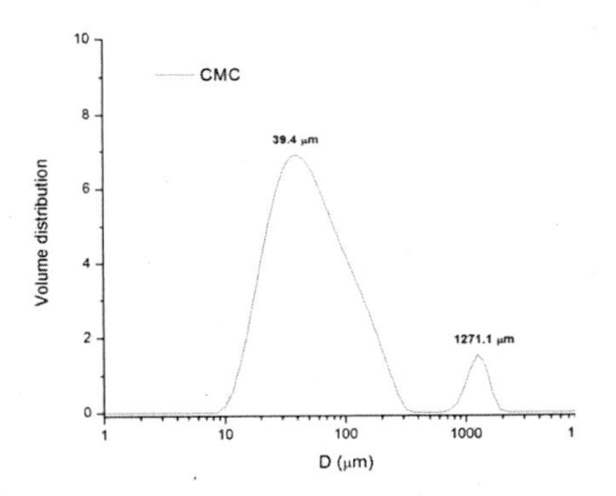

Figure 3. Particle size distribution of solid CMC.

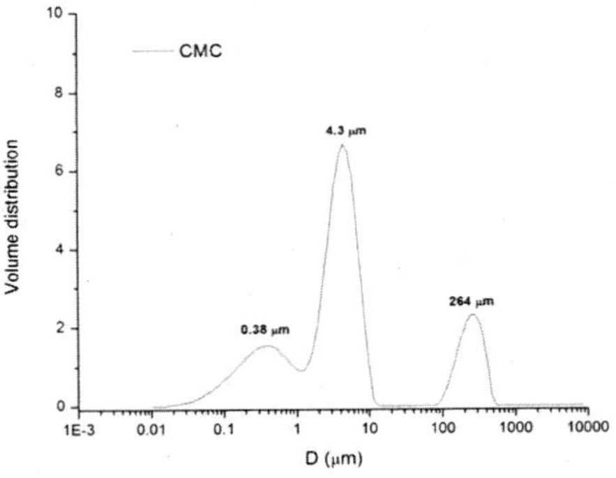

Figure 4. Particle size distribution of a suspension of CMC.

Figure 2 presents particle size volume distribution of the as-received dry HPMC. The particle size in volume is monomodal, but the curve has a very broad shape, with a maximum at ~267 µm, the broadness actually suggesting that the powder may be constituted of aggregates of different sizes.

As shown in Figure 3, the particle size volume distribution of CMC (dry powder) is bimodal, displaying large aggregates of ~39 µm along with few very large ones, ~1270 µm in size. A suspension of CMC in water displays a different particle size distribution, trimodal, implying that in the presence of water, larger aggregates are breaking down into mainly ~4 µm, along with submicron (~0.4 µm) and larger (~260 µm). By comparing data in Figures 3 and 4, it appears that by adding the CMC to water prior to addition to the TiO_2 powder is beneficial for the binder dispersion in the final paste.

Two pastes prepared with TiO_2 and the two binders, HPMC and CMC, respectively, have been subjected to capillary rheometry, composition of the two pastes being shown in Table 1, along with the calculated extrusion parameters. A simple model was employed for the characterization of the flow properties of the ceramic pastes[23], based on the extrusion of pastes through a using cylindrical barrel and die by a piston extruder. The relation between the pressure P applied in the barrel and the extrudate velocity V during flow from a cylindrical barrel of diameter D_o into a die-land of diameter D and length L could be described by the equation[23]:

$$P = P_1 + P_2 = 2(\sigma_0 + \alpha V)\ln(D_0 / D) + 4(\tau_0 + \beta V)(L / D) \qquad (1)$$

where $(\sigma_0 + \alpha V)$ is termed the paste yield value, σ_0 is the bulk yield stress extrapolated to zero velocity, α is a factor characterizing the effect of velocity.

The first term in Equation (1) is related to flow into the die from the barrel, implying that the flow is plastic. If we do account for the plasticity, for a circular barrel and a circular land, the first term in equation 1 can be written simply as follows:

$$P_1 = 2\sigma \ln(D_0 / D) \qquad (2)$$

But generally, in paste flows, viscous effects are in operation, with the stresses generated being related to the rate of strain, therefore the first term is more accurately described as $(\sigma_0 + \alpha V)$. If we recall the definition of the viscosity as the ratio of shear stress to the shear rate,

$$d\gamma / dt = \dot{\gamma} \qquad (3),$$

that is further

$$\eta = \tau / \dot{\gamma} \qquad (4),$$

we can consider αV as analogous to the product $\eta \dot{\gamma}$ in liquid shear flow. The rate of extension is proportional to V and thus $(\sigma_0 + \alpha V)$ can be regarded as the bulk yield stress corrected for shear rate.

The second term in Equation (1) describes the flow along the bore of the die land. If a motion of the paste in the die land is considered, it is found that the bulk of material moves at a constant velocity with only a very thin layer near the wall subject to shear strain. The pressure drop P_2 in Equation (1) generates a net force on the paste which is opposed by the wall shear force. In the thin layer of liquid next to the wall of the die in which shearing takes place, viscous effects are significant. Parameters τ_0 and β are describing the paste, τ_0 is the wall shear stress extrapolated to zero velocity, while β is a factor which accounts for the velocity dependence of the wall shear stress and is termed the wall velocity factor. The four characterizing extrusion parameters σ_0, α, τ_0 and β (Table 1) have thus been introduced.

Consequently, the static characteristics of the paste, σ_0 and τ_0 can be determined by extrapolation from results in which extrudate velocities have been varied (1mm/min and 10mm/min, respectively) and dies with different ratios L/D have been employed (16 and 32)[23].

Table I. Paste composition and calculated extrusion parameters.

Constant/ Paste	Composition (wt%)	Bulk Stress (MPa) σ_0	Velocity Factor, Barrel region (MPa min/mm) α	Wall Stress (MPa) τ_0	Velocity Factor, Die Wall (MPa min/mm) β
TiO_2- CMC	68% TiO_2 4% CMC 28% water	0.112	0.242	0.0005	0.015
TiO_2- HPMC	68% TiO_2 2% HPMC 30% water	0.137	0.05	0.013	0.0034

Observing the results shown in Table 1, the addition of a low viscosity soluble polymer (HPMC) translates into an only slightly higher bulk yield stress (0.137 MPa), but in a much higher wall stress ($1.3x10^{-2}$ MPa), compared with the corresponding values obtained using the low-soluble CMC additive (0.112 MPa, $5x10^{-4}$ MPa, respectively). When using a more viscous liquid phase (HPMC solution) in paste preparation, it is expected that the term $(\sigma_0 + \alpha V)$ to become more velocity dependant, therefore α, the bulk velocity factor value to be higher. The bulk velocity factor for the TiO_2-HPMC paste (0.05 MPa min/mm) was much lower than that for the TiO_2-CMC paste (0.242 MPa min/mm). Therefore, the bulk yield stress variation with shear rate described by $(\sigma_0 + \alpha V)$ for TiO_2-HPMC paste will be allowing its extrusion on a wider range of velocities V than for the TiO_2-CMC paste, when the term $(\sigma_0 + \alpha V)$ will increase significantly with the extrusion velocity. By using a low solubility organic additive, CMC, the wall stress extrapolated to zero velocity (τ_0) is much smaller than that obtained when a soluble organic additive, HPMC, was used, while the β velocity factor is displaying an opposite trend ($1.5x10^{-2}$ MPa min/mm for CMC and $3.4x10^{-3}$ MPa min/mm for HPMC).

The above analysis considers the extension of the bulk material during its passage from the barrel into the die land followed by shear of the material against the wall of the die land, with the strain being confined to a narrow region adjacent to the wall. It is essential that the paste composition should be such there will be sufficient liquid at the wall to achieve lubrication. Values of β suggest better flow characteristics over a wide range of V for the case when HPMC was used, which implies that enough liquid phase exists in the TiO_2-HPMC paste, ensuring its easy extrusion. A very important property of soluble polymers (such as HPMC) is that their aqueous solutions have a remarkable water-holding capacity[6]. The greater the water retention and the less free water present in the system, the easier the flow through the die can be expected. Thus, as β parameter value suggests for the TiO_2-HPMC paste, easier flow was achieved for this paste as compared to the TiO_2-CMC paste. The low solubility of CMC in water makes its behavior within the batch closer to an inorganic additive and leaves a higher free water content in the paste, which makes flow through the die increasingly difficult at higher velocities, V.

In our attempt to correlate extrusion characteristics with morphological characteristics of the extrudates, especially particle packing, along with assessing the influence of organic additives addition, particle size distribution, surface analysis and SEM analysis was performed.

Figure 5 displays particle size distribution analysis of extrudates obtained using CMC and HPMC, along with ones without any organic additive and TiO_2 pristine powder. All measurements have been recorded on suspensions, to assess the aggregates size distribution in suspension before and after extrusion of the Titania powder, in the presence and absence of an organic additive.

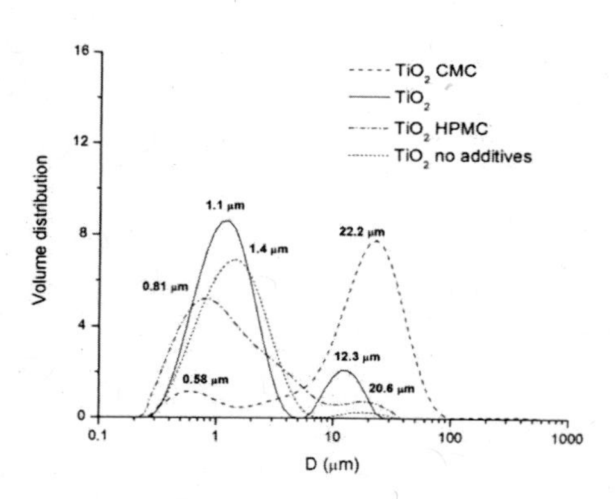

Figure 5. Particle size distribution of TiO_2extrudates and TiO_2.

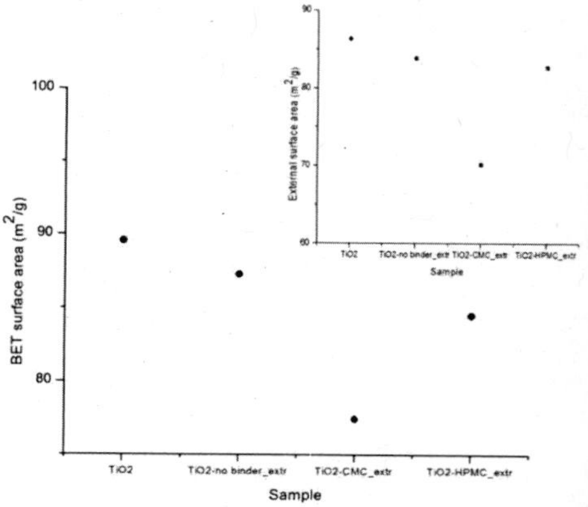

Figure 6. Variation of BET and external (insert) surface area of extrudates and TiO_2.

The particle size volume distribution is bimodal for the pristine powder and all the obtained extrudates. It is observed that a suspension of the pristine powder exhibits mainly aggregates of 1.1 μm along with fewer large ones, of 12.3 μm. Measurements taken on suspensions of extrudates display that binder additions lead to decrease in the size of the smaller aggregates (0.58μm for CMC and 0.81μm for HPMC) and increase in size of the larger ones (22.2 μm for CMC and 20.6 μm for HPMC). Extrudates obtained without any organic additives showed almost no change in the size of small aggregates (1.4 μm), but they displayed larger ones of the same size as the HPMC extrudates (20.6 μm). According to Figure 5, extrusion leads to different aggregate size when a soluble or an insoluble organic additive has been employed.

To assess the extrudates surface properties and compare them with the pristine TiO_2 powder, surface area determination, along with pore volume and area distribution by BJH method was carried out.

As presented in Figure 6, the BET surface area of the extrudates varies slightly as compared to the pristine TiO_2, while TiO_2-CMC extrudates presented the lowest value. The surface area of the pristine TiO_2 accounts mostly as external surface area and this is also preserved in the extrudates. The line of demarcation between the two kinds of surface is sometimes done in an arbitrary way, but the external surface may perhaps be taken to include all the prominences and all of the cracks which are wider than they are deep.

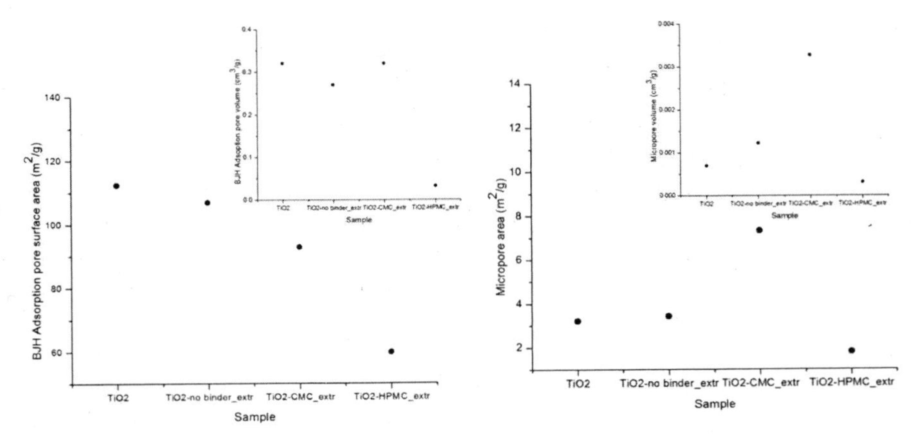

Figure 7. Variation of BJH pore area and volume (insert) of extrudates and TiO_2.

Figure 8. Variation of micropore area and volume (insert) of extrudates and TiO_2.

Conversely, the internal surface will comprise the walls of all cracks, pores and cavities which are deeper than they are wide[24]. Among the solids with a surface area which is wholly external in nature are those usually constituted by aggregates of rather spherical particles, thus particle-particle contacts will be very small in area. TiO_2 used in this study has these morphological characteristics, as it will be shown further. The interparticulate junctions will be rather weak and many of them will become broken apart by mechanical treatments.

Micropore area and volume vary slightly from the pristine titania to the extrudates, being almost unchanged in the extrudates prepared without organic additives and slightly lower in the extrudates prepared with HPMC (Figure 8). Microporosity increased slightly in the extrudates prepared with CMC. One of the origins of micropores is imperfect packing arrangements of the bulk material producing a lack of crystalline alignment[24]. Figure 7 presents the variation of the cumulative pore area and pore volume of pores with diameters between 0.85 and 150 nm, calculated by the BJH method[25]. Pore area decreases all the way from the pristine titania powder, extrudates with no organic additive, followed by TiO_2-CMC extrudates, while the TiO_2-HPMC extrudates had the smallest value. Pores in the range of 0.85 and 150 nm diameter take into account micropores (less than 2 nm), mesopores (2-50 nm) and macropores (> 50 nm). Mesopores correlate with the space between agglomerated particles and serve as passages, providing a transport system, to the micropores[24]. Major lattice structure defects, such as fissures, or racks within a solid lead to the formation of macropores, which may be treated as an open surface. The macropores along with the mesopores act as transport pores, allowing access to the internal surface (when applicable) and microporosity. The macropores do not contribute considerably to the surface area, typically less than 2 m^2/g. As seen in Figure 7, shear applied to the TiO_2-no additives paste led to a decrease in the pore area in the corresponding extrudates as compared to the pristine titania. Since according to Figure 8, upon adding HPMC, the volume and the area of the micropores in the TiO_2-HPMC extrudates did not change notably, the significant decrease in the pore area and volume in Figure 7 accounts basically for the mesopores. Decreasing in the mesopore volume in the TiO_2-HPMC extrudates accounts for decreasing significantly of the volume between the agglomerated particles, this concludes that HPMC acts primarily as an organic binder and compacts the agglomerated solid particles in the final material. The pore area of the TiO_2-CMC extrudates is higher than that for TiO_2-HPMC extrudates, along with a higher pore volume, the same as for the pristine powder, accounting for a less compacted extrudate.

To visualize the aggregate microstructure, and further assess the packing of the particles after extrusion, FESEM images of the solid as-prepared extrudates have been taken and presented in Figure 9.

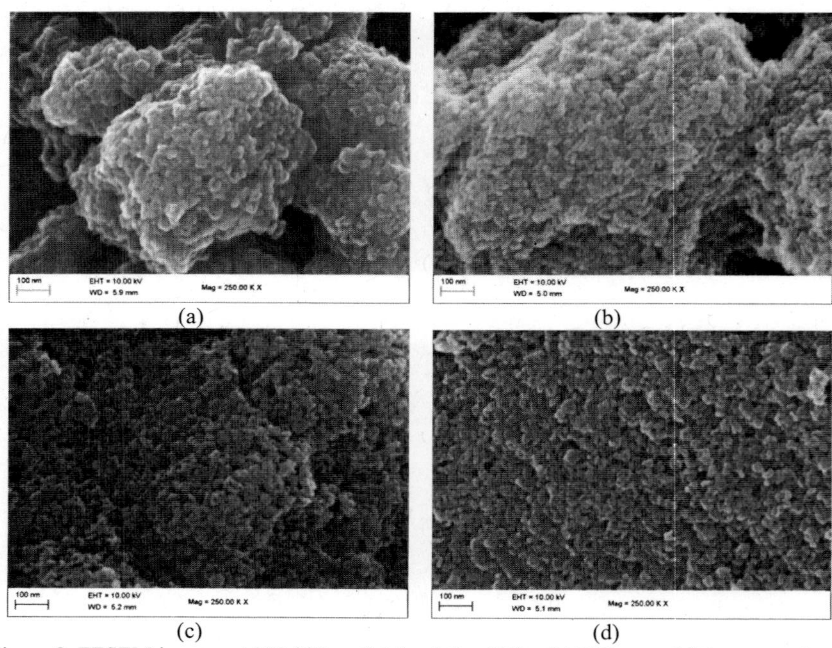

Figure 9. FESEM images at 250,000 x of: (a) pristine TiO_2; (b) TiO_2-no additives extrudates; (c) TiO_2-CMC extrudates and (d) TiO_2-HPMC extrudates.

The pristine TiO_2 powder (Figure 9a) is constituted by aggregates of 0.5~1μm, in turn created by ~ 20 nm primary particles. By applying shear on a paste constituted only by TiO_2 and water, the aggregated structure of the pristine TiO_2 changed and primary particles appear to be slightly more rigidly joined together in larger agglomerates, of ~2μm (Figure 9b). The TiO_2-CMC extrudates (Figure 9c) are constituted by close-packed primary particles into aggregates (<1 μm) rigidly joined together in larger agglomerates. Microstructure of the TiO_2-HPMC extrudates shows quite homogeneously close packed primary particles into one large agglomerate (Figure 9d). A good correlation between the surface and pore area variation with the agglomerate and packing structure of the extrudates and pristine powder was found. The lowest mesoporosity was identified for TiO_2-HPMC extrudates, which at the same time presented the highest compaction of the primary particles of titania. This could be explained by specific interactions of HPMC with the powder. When a solution containing a soluble polymer (HPMC) is added to a powder, the polymer chain, which is usually about the size of 0.1-1μm attaches to several particles causing a so-called "flocculation" effect[14]. By adsorption of HPMC, the electrical repulsion between titania particles is reduced and causes them to be held together by van der Waals forces. It was found that the stronger the binding forces, the larger can the aggregates grow under given shear conditions. Polymeric flocculants provide many links between particles, several "bridges" between particles are generally formed and therefore large strong flocs are produced. Since the TiO_2 employed contains residual sulfate (1.8% SO_3 wt%), the ionic strength is quite high, the range of electrical repulsion is reduced, therefore a lower molecular weight polymer, such as the HPMC presently employed, was proven to be effective in creation of homogeneous extrudates. For such polymers, the concept of bridging through a repulsion barrier is supported by a number of experiments[14].

By mixing a suspension of CMC with the titania powder, it is expected that this makes the powder extrudable by filling voids between the aggregates (most of the suspended CMC particles were small, around 40 nm), so providing a lubricating layer around them. This is consistent with the paste rheology and also with the lowest surface area value found for TiO_2-CMC extrudates. Interaction between CMC and TiO_2 powder leads to change in aggregation and morphology of the corresponding extrudates. Applying shear to a paste composed only by TiO_2 and water, causes the aggregates to break down and re-arrange into larger ones, but with a more compact packing of the primary 20 nm particles than in the pristine powder, but the general aggregation features are preserved.

SUMMARY

Extrusion behavior of TiO_2-based pastes prepared with two different organic additives along with the microstructure and porosity of the corresponding extrudates have been studied. The use of water soluble (HPMC) versus a low soluble (CMC) organic additive was compared.

The use of a soluble cellulose polymer with low viscosity was found beneficial for the overall extrusion process, an important factor being its remarkable water-holding capacity. Capillary data predicted that such a polymer allows extrusion in a broad range of applied shear, without changing much the process parameters, and improves extrudates shape and compaction. A good correlation between the surface and pore area variation with the agglomerate and packing structure of the extrudates and pristine TiO_2 powder was found. Decreased pore volume in the TiO_2-HPMC extrudates led to the conclusion that HPMC acts by compacting the agglomerated solid particles in the final material. Microstructure of the TiO_2-HPMC extrudates shows quite homogeneously packed primary particles into a large agglomerate. Conversely, the increase in pore area of the TiO_2-CMC extrudates accounted for a less compacted extrudate.

The above results have been explained in view of the specific interactions of the different organic additives at the liquid-solid interface. The use of water soluble additive versus a water low soluble one has lead to different morphologies and aggregation characteristics of TiO_2-based extrudates. In absence of an additive during extrusion, TiO_2 aggregates are destroyed and the primary particles re-arranged under shear, but the specific aggregation features of the pristine powders are preserved.

ACKNOWLEDGEMENTS

The authors would like to acknowledge the financial support provided by the Ceramic and Composite and Optical Materials Research Center at Rutgers University, an NSF/IUCRC. In addition, the authors would like to thank Malvern Instruments, Ltd. for providing the instrumentation used in the testing.

REFERENCES

[1]N. Phonthammachai, T. Chairassameewong, E. Gulari, A.M. Jamieson, S. Wongkasemjit, Structural and rheological aspect of mesoporous nanocrystalline TiO_2 synthesised via sol-gel process, *Microporous Mesosporous Mater* **66**, 261-171 (2003).
[2]A. Teleki, R. Wengeler, L. Wengeler, H. Nirschl, S.E. Pratsinis, Distinguishing between aggregates and agglomerates of flame-made TiO_2 by high-pressure dispersion, *Powder Technology* **181**, 292-300 (2008).
[3]W. J. Tseng, K. Lin, Rheology and colloidal structure of aqueous TiO_2 nanoparticle suspensions, *Materials Sci. and Engn.* A **355**, 186-192 (2003).
[4]H.K. Kammler, L. Madler, S.E. Pratsinis, Flame synthesis of nanoparticles, *Chemical Engineering and Technology* **24(6)**, 583-596 (2001).
[5]P. Forzatti, C. Orsenigo, D. Ballardini, F. Berti, On the relations between the rheology of TiO_2-based ceramic pastes and the morphological and mechanical properties of the extruded catalysts, *Preparation of Catalysts VII*, Ed. Delmon et al, Elsevier, (1998).
[6]A. Ekonomakou, A. Tsetsekou, C.J. Stournaras, The influence of Binder Properties on the Plasticity and the Properties of Raw Extruded Ceramics, *Key Engineering Materials*, **132-136**, 420-423 (1997).
[7]L. Chevalier, E. Hammond, A. Poitou, Extrusion of TiO_2 ceramic powder paste, *J. Mater. Proc. Technol.* **72**, 243-248 (1997).
[8]P. Forzatti, D. Ballardini, L. Sighicelli, Preparation and characterization of extruded monolithic ceramic catalysts, *Catal. Today* **41** 87-94 (1998).
[9]M. Miller, R.A. Haber, Use of Montmorillonites as Extrusion Aids for Alumina, *Ceram. Eng. Sci. Proc.* **12**, 33-48 (1991).
[10]I.M. Lanchman, J.L. Williams, Extruded monolithic catalyst supports, *Catal. Today*, **14**, 317-329 (1992)
[11]D. Ballardini, L. Sighicelli, C. Orsenigo, L. Visconti, E. Tronconi, P. Forzatti, A. Bahamonde, E. Atanes, J.P. Gomez Martin, F. Bregani, *Studies in Surface Science and Catalysis*, Ed. by J.W. Hightower et al. **101**, 1359 (1996).
[12]K. Hayakawa, S. Niinobe, *Extrusion or injection molding of ceramic, glass, or polymer powders in water-soluble cellulose ether binder*, U.S. Pat. Appl. Publ. US 2007293387 (2007).
[13]R. Bayer, M. Knarr, *Cellulose ether additives for the extrusion of ceramic masses*, Ger. Offen., Chemical Indexing Equivalent to 149:85382 (WO) (2008).
[14]Th.F. Tadros, *Solid/Liquid Dispersions*, Academic Press, London (1987).
[15]E. Ott, *High polymers, Volume 5: Cellulose and Cellulose derivatives*, Interscience Publishers, Inc., New York (1946).
[16]R. Bayer, *Additives based on cellulose ethers, defoamers, and plasticizers for extrusion of porous ceramics,* PCT Int. Appl. WO 2009153617 (2009).

[17]R.J. Huzzard, S. Blackburn, A Water-Based System for Ceramic Injection Moulding, *J. Eur. Ceram. Soc* **17** 211-216 (1997).
[18]Grayson, M. *Encyclopaedia of Chemical Technology*; John Wiley & Sons: New York, **5**, (1979).
[19]R.L. Whistler, J.N. BeMiller, *Industrial Gums, Polysaccharides and Their Derivatives*, 2nd ed.; Academic Press: London, (1973).
[20]H.C.W. Foerst, H. Buchholz-Meisenheimer, *Ullmans Encyklopadie der technischen Chemie*; Urban & Schwarzenberg: Berlin, (1970).
[21]AQUALON CO, U.S. Patent US 4948-430-A (1990).
[22]W. Gleissle, J. Graczyk, H. Buggisch, *KONA Powder and Particles*, **11** 125- (1993).
[23]J. Benbow and J. Bridgwater, *Paste flow and extrusion*, Clarendon Press, Oxford, (1993).
[24]S.J. Gregg, K.S.W. Sing, *Adsorption, Surface area and Porosity*, Second Ed, Academic Press, London (1982).
[25]E.P. Barrett, L.G. Joyner and P.P. Halenda, The Determination of Pore Volume and Area Distributions in Porous Substances. I. Computations from Nitrogen Isotherms, *J. Am. Chem. Soc.* **73**, 373-380 (1951).

PHASE TRANSITION AND CONSOLIDATION OF COLLOIDAL NANOPARTICLES

Yoshihiro Hirata, Naoki Matsunaga and Soichiro Sameshima
Department of Chemistry, Biotechnology, and Chemical Engineering, Kagoshima University, 1-21-40 Korimoto, Kagoshima 890-0065, Japan

ABSTRACT

The activity and chemical potential of dispersed and flocculated particles of one-component colloidal system were defined. The combination of thermodynamics of colloidal suspensions and the DLVO theory succeeded in constructing the colloidal phase diagram in a map of surface potential and solid content of particles of 10-1000 nm diameters. This phase diagram can predict the dependence of packing density on particle size. The experimentally-determined packing density agreed with the prediction from the phase diagram. The consolidation behavior of nanometer-sized particles at 20-800 nm was examined using a pressure filtration apparatus. A phase transition from a dispersed suspension to a flocculated suspension occurred when applied pressure exceeded a critical value. Based on the colloidal phase transition, new filtration theories were developed for a flocculated suspension. A good agreement was shown between the developed theories and experimental results.

INTRODUCTION

Powder processing through colloidal suspensions is widely recognized to be an effective forming method which can improve a microstructure and resultant physicochemical properties of advanced ceramics[1–4]. The guideline of colloidal processing is interpreted by the interaction energy between charged particles, which corresponds to the summation of van der Waals attraction energy and the electrostatic repulsive energy due to the electrical double layer (DLVO theory)[5, 6]. The DLVO theory explains the dispersed and flocculated states depending on the surface potential of the particles. However, we need a description of the thermodynamics of colloidal systems to understand the particle size effect on the stability of a colloidal suspension[7–9]. In our previous paper[9], we studied the thermodynamics of a colloidal suspension containing dispersed and flocculated particles. From the derived thermodynamic relations, we have constructed a one-component colloidal phase diagram as functions of surface potential and solid content of particles. In this paper, the properties of a colloidal suspension predicted from the constructed phase diagram are compared with the experimentally measured results. For the consolidation of nanometer-sized particles by pressure filtration, it was found that deviation between the established filtration theory[10] for well dispersed particles and experimental result for the pressure drop across the consolidated layer increased when the applied pressure exceeded a critical value. This deviation is related to the phase transition from dispersed to flocculated suspension[11]. The important factors affecting the phase transition of colloidal suspension are pointed out in this paper from the previously measured experimental results. Furthermore, a new filtration theory was developed for the consolidation of a flocculated suspension after the phase transition[11]. This theory was evaluated with the experimental results of pressure filtration of nanometer-sized SiC particles under a constant applied pressure or a constant crosshead speed of piston.

COLLOIDAL PHASE DIAGRAM

Figure 1 shows a structure model of colloidal suspension containing both dispersed and flocculated particles. It is well known that highly charged particles form a well dispersed state due to the strong repulsive energy which gives a low viscosity suspension and high packing density after consolidation. On the other hand, the colloidal particles at isoelectric point are connected by van der Waals attractive energy. The flocculated particles provide a high viscosity suspension and form a porous green compact by filtration through gypsum mold. The activity of dispersed particles (a_d) is expressed by Henry's law and equal to the product of the molar fraction (α) of dispersed particles in a suspension containing both dispersed and flocculated particles and the activity coefficient (γ^0) expressed by C_0/C_{max} (C_0 : total volume fraction of dispersed and flocculated particles, C_{max} : maximum packing density of particles). For the close packing and random close packing models, C_{max} is 0.740 and 0.637, respectively. The γ^0 value results in 1 for $C_0 = C_{max}$. That is, a_d is defined as a ratio of volume of dispersed particles (C_d) to the maximum volume packed, $a_d = C_0\alpha/C_{max} = C_d/C_{max}$. The activity of flocculated particles (a_g) follows Raoult's law and is expressed as $(1-\alpha)$ using the Gibbs–Duhem equation. The chemical potential (μ) is represented by the defined activity. The difference of chemical potential of both the particles is expressed by Eq.(1).

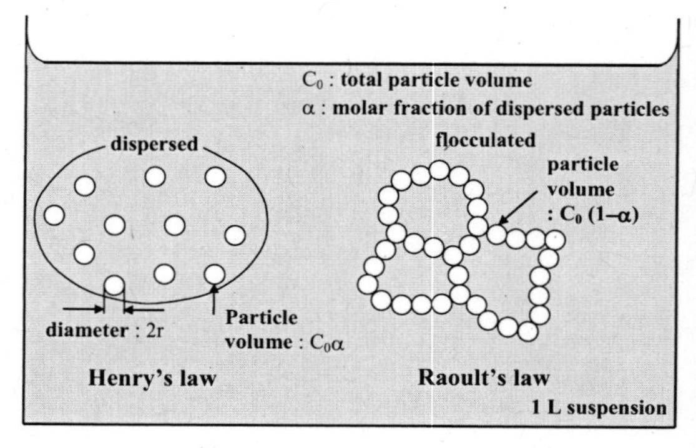

Figure 1. A suspension model containing dispersed particles and flocculated particles.

$$\Delta\mu = \mu_g - \mu_d = \Delta\mu_0 + RT\,ln\left(\frac{1-\alpha}{\gamma^0\alpha}\right) \tag{1}$$

where $\Delta\mu_0$ represents $\mu_{g0}-\mu_{d0}$, and the μ_{g0} and μ_{d0} correspond to the chemical potential of flocculated and dispersed particles for $\alpha = 0$ and 1, respectively. The condition of $\Delta\mu > 0$ indicates the change of flocculated particles to dispersed particles. On the other hand, dispersed particles change to flocculated particles at $\Delta\mu < 0$. The dispersed and flocculated states reach an equilibrium at $\Delta\mu = 0$. The dispersed particles in a suspension collide to form flocculated particles. The collision rate is usually expressed by second order reaction[12, 13]. That is, the fraction (α) of dispersed particles, activity and chemical potential of dispersed or flocculated particles in a suspension change with settling time. The above relation leads to Eq. (2),

$$\Delta\mu = RT\,ln\left(\frac{t}{t_{c1}}\right) \tag{2}$$

The $\Delta\mu$ value becomes $\Delta\mu < 0$ at $t < t_{c1}$ and $\Delta\mu = 0$ at $t = t_{c1}$ (critical time). No change of $\Delta\mu$ occurs at $t > t_{c1}$, because the equilibrium state is achieved. The time t_{c1} required to the equilibrium state is determined from Eq.(3),

$$t_{c1} = \frac{\gamma^0}{n_i k_1} exp\left(-\frac{\Delta\mu_0}{RT}\right) \tag{3}$$

where n_i is the total particle number in a suspension and k_1 the rate constant for the flocculation. The rate constant is expressed by Arrhenius equation Eq. (4)[12, 14],

$$k_1 = k_{10}\, exp\left(-\frac{\Delta G_{m1}}{RT}\right) \tag{4}$$

where k_{10} is the frequency factor and ΔG_{m1} represents the activation energy for the migration of dispersed particle to another dispersed particle or to adjacent flocculated particle. The critical time t_{c1} for the phase transition from dispersed to flocculated state is expressed as follows by the combination of Eqs. (3) and (4),

$$t_{c1} = \frac{\gamma^0}{n_i k_{10}} exp\left(-\frac{\Delta\mu_0 - \Delta G_{m1}}{RT}\right) \tag{5}$$

Figure 2 shows the energy difference of dispersed and flocculated particles at $t < t_{c1}$ and $t = t_{c1}$. The dispersed particles form a particle cluster when they have an energy higher than the potential barrier, ΔG_{m1}. Equation (5) is transformed to Eq. (6) using the basic relations of $\gamma_0/n_i = 4\pi r^3/3C_{max}$ (r: particle radius), $\Delta\mu_0 = \Delta H_0 - T\Delta S_0$ and $\Delta G_{m1} = \Delta H_{m1} - T\Delta S_{m1}$ (H : enthalpy, S : entropy, T : temperature). The entropy difference (ΔS_0) of dispersed and flocculated particles at $a_d = a_g = 1$ should be close to ΔS_{m1} of dispersed and flocculated particles present at t = t, leading to $\Delta S_{m1} - \Delta S_0 \approx 0$.

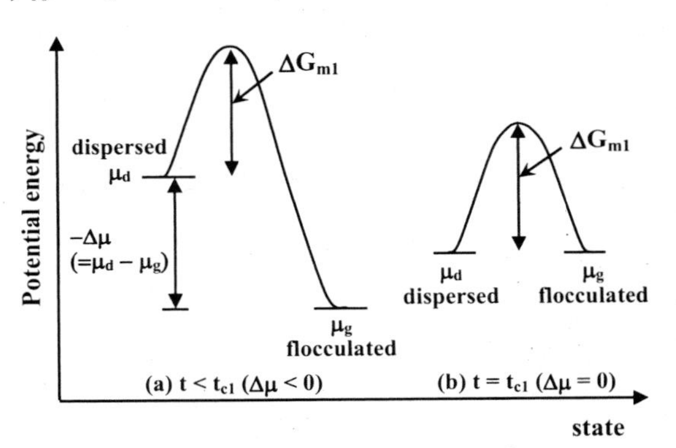

Figure 2. Energy difference of dispersed and flocculated particles at (a) $t < t_{c1}$ (critical time for equilibrium) and (b) $t = t_{c1}$. ΔG_{m1} represents the activation energy for one particle to migrate to another particle.

$$\ln t_{c1} \approx A + 3 \ln r + \frac{\Delta H_{m1}}{RT} \tag{6}$$

A is the constant value $(\ln(4\pi/3C_{max}k_{10}) - \Delta H_0/RT)$. Equation (6) indicates that (i) t_{c1} depends on particle size and ΔH_{m1} (enthalpy for migration of dispersed particle) and (ii) t_{c1} becomes longer for a larger r and for a higher ΔH_{m1} value. Equations (2) and (6) were coupled to give Eq. (7) for a fresh well dispersed suspension ($\ln t << \ln t_{c1}$).

$$\frac{\Delta\mu}{RT} \approx -\ln t_{cl}$$
$$\approx -A - \frac{\Delta H_{ml}}{RT} - 3\ln r < 0 \quad (7)$$

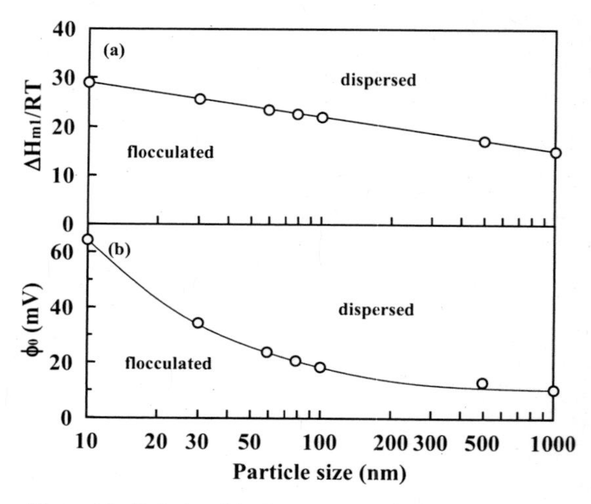

Figure 3. Relationship between particle size and (a) enthalpy (ΔH_{ml}) for the migration of dispersed particles to form particle cluster, and (b) surface potential corresponding to ΔH_{ml}.

The enthalpy term is equivalent to the maximum value of the interaction energy (E_i(max)) as a function of distance between two particles in the DLVO theory and depends on suspension temperature[6, 9]. The activation energy (ΔH_{ml}) to prevent the forming of particle cluster by collision of dispersed particles is a function of particle size and should increase when particle size becomes small ($\Delta H_m/RT > -A-3\ln r$). However, it is difficult to determine the theoretical A value in Eq. (7), because unknown two parameters k_{10} and ΔH_0 are included in A value. As a starting point, an experienced criterion of ΔH_{ml}(critical) / RT $\approx$ 15 (for $\Delta\mu/RT = 0$) was assumed for a particle of 1 μm diameter[6]. The E_i(max) was calculated for many surface potential φ mV under the constant conditions of electrical double layer thickness 100 nm, Hamaker constant 10.7×10^{-20} J, atomic valence of electrolyte in a suspension +1 and temperature 298 K. The detailed E_a and E_r equations are shown in our previous paper[6].

Figure 3(a) shows the relation between $\Delta H_{ml}/RT$ value and ln r for Eq. (7). A large ΔH_{ml} provides a high stability of colloidal suspension. The critical ΔH_{ml} for $\Delta\mu/RT = 0$ increases linearly with decreasing ln r. Figure 3(b) shows the critical surface potential (ϕ_0) corresponding to $\Delta H_{ml}/RT$ in Fig. 3(a). The ϕ_0 value increases drastically when particle size becomes smaller than 100 nm. On the other hand, the distance between two particles (H(max)) corresponding to E_i(max) was

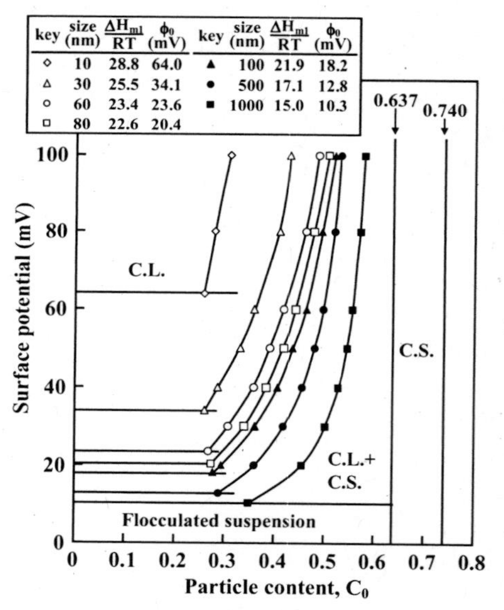

Figure 4. Comparison of colloidal phase diagrams for particle size range from 1000 nm to 10 nm. C.L. : colloidal liquid, C.S. : colloidal solid

converted to the initial concentration of colloidal particles C_0, using Eq.(8) for random close packing model[15, 16].

$$H(max) = 2r\left[\left(\frac{1}{3\pi C_0}+\frac{5}{6}\right)^{\frac{1}{2}} - 1\right] \quad (8)$$

The H(max) value approaches 0 at C_0 = 63.7 vol%. Two dispersed particles within the distance of $0 < H < H(max)$ have a strong tendency to make a particle cluster because of the rapid decrease of E_i to $-\infty$ with decreasing H value. Based on the above analysis, colloidal phase diagrams for one-component systems of 10–1000 nm diameters were constructed. Figure 4 compares the colloidal phase diagrams for the particle size range from 1000 nm to 10 nm. The phase transition from a colloidal liquid to a flocculated suspension containing 1000 nm particles occurs at ϕ = 10.3 mV (Fig. 3(b)). The formed flocculated suspension consists of two phases of flocculated colloidal particles and an aqueous solution. The critical surface potential ϕ_0 increases with decreasing particle size. On the other hand, the C_0 range for the mixture of colloidal liquid and colloidal solid becomes narrow at a higher ϕ value (Eq. (8)). The C_0 range for the mixture of colloidal liquid plus colloidal solid becomes wide with decreasing particle size. In this C_0 range, it is difficult to change the position of particles, because the distance between two particles is substantially shorter than the particle size. The above comparison suggests the consolidation of fine particles results in a low packing density.

Figure 5 shows the packing density determined from the liquidus line at 100 mV and at ϕ_0 mV (solidius lines) of surface potential as a function of particle size. The packing density of the dispersed particles, decreases from 58 % at 1000 nm to 30 % at 10 nm. However, the packing density at $\phi = \phi_0$ is not sensitive to the particle size and is in the range from 26 to 35 vol%. The difference of packing density at 100 mV and ϕ_0 becomes smaller with decreasing particle size. Figure 5 also shows the packing density of ceramic particles with sizes of 24–800 nm. In the experiment, colloidal suspensions at 20–48 mV of

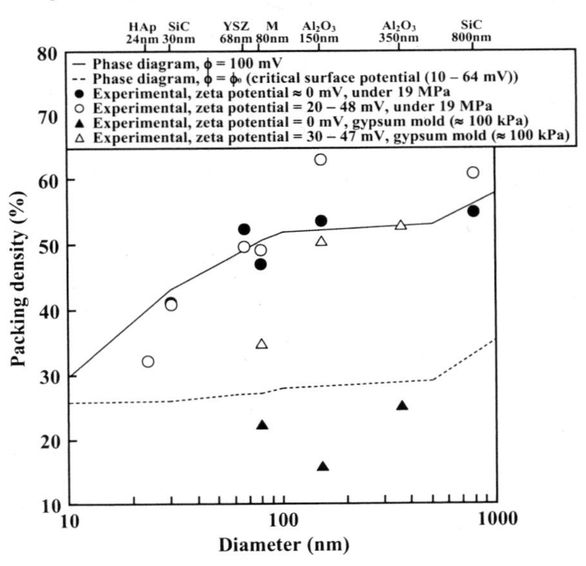

Figure 5. Packing density of nanometer-sized ceramic particles as a function of particle size. YSZ, HAp and M represent yttria-stabilized zirconia, hydroxyapatite and mullite, respectively.

zeta potential and at near the isoelectric point (flocculated particles) were consolidated by a pressure filtration apparatus under a pressure of 19 MPa or using a gypsum mold (suction pressure: 50–100 kPa). The packing densities by pressure filtration are very close to the compositions of liquidus lines at 100 mV of surface potential. The surface potential of the particles smaller than 100 nm does not affect the packing density. This result is explained by the shift of the solidus line to a higher ϕ with decreasing particle size. The colloidal suspensions for the particles of 20–70 nm were in the flocculated state at given zeta potentials. In the submicrometer range, the packing density was influenced by the surface potential and became higher than the compositions of liquidus lines. This result reflects the crystallization of highly dispersed particles during the consolidation. On the other hand, the low packing density of 16–26 vol% was measured for 80–350 nm particles at 0 mV of zeta potential under a low suction pressure. This result is well explained by the density curve determined from the solidus line of the phase diagram. Apparently, the flocculated particles can be consolidated to a high packing density predicted from the phase diagram under a high consolidation pressure.

PHASE TRANSITION OF COLLOIDAL SUSPENSION UNDER APPLIED PRESSURE

As seen in Fig. 5, a good agreement was seen in the packing density between the pressure filtration and colloidal phase diagram, indicating the colloidal nanoparticles were randomly packed. According to the phase diagrams in Fig. 4, the flocculated particles with low zeta potential < ϕ_0 can be packed toward the density of liquidus line under the applied pressure. The consolidation behavior of colloidal particles with high and low zeta potential was studied in our previous papers[11, 17]. Typical results during the consolidation of Al_2O_3 suspensions at pH 7.0–7.2 are shown in Fig. 6[18]. The positively charged alumina particles (sample a, ≈ 40 mV, 0.2 μm diameter) were filtrated under a constant crosshead speed of piston in accordance with the established filtration model for well dispersed particles : ΔP_t (applied pressure) $\propto V_f$ (volume of filtration). However, an almost plateau region of ΔP_t appeared in region III. This change in the consolidation behavior is related to the phase transition from a dispersed suspension to a flocculated suspension at a critical applied pressure (ΔP_{tc}, phase transition pressure). The factors affecting ΔP_{tc}

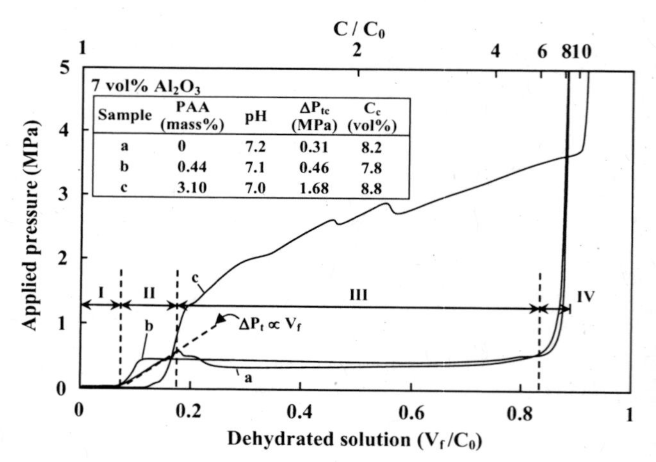

Figure 6. Relationship between normalized filtration volume (V_f / C_0) and applied pressure for Al_2O_3 suspensions at pH 7.0–7.2 (C_0 : initial suspension volume). The four regions indicated correspond to sample (a) without polyacrylic ammonium. The value C_c represents the solid content of the suspension at P_{tc} of applied pressure.

are the zeta potential, concentration of suspension, size of particles, polyelectrolyte dispersant and electric field[11, 18].

Figure 7 shows ΔP_{tc} of several nanometer-sized particles as a function of the zeta potential[11]. Increase of zeta potential, decrease of particle concentration and increase of particle size shift ΔP_{tc} to a high pressure. The ΔP_{tc} is also influenced by the dissociation and amount of polyelectrolyte added[18]. The addition of a large amount of highly charged polyacrylic ammonium suppressed the phase transition from dispersed to flocculated state of alumina particles at a given applied pressure (Fig. 6, sample c). When an alternating electric field was applied, the ΔP_{tc} value decreased drastically. The flocculation of a suspension was accelerated by the electric filed during pressure filtration.

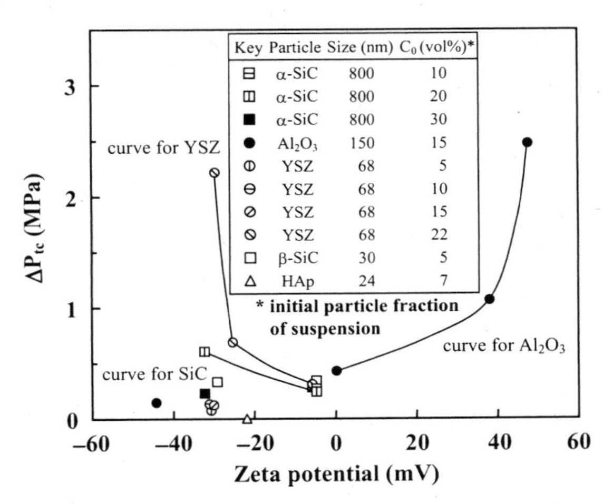

Figure 7. Phase transition pressure and zeta potential for several nanometer-sized particles. YSZ and HAp represent yttria-stabilized zirconia and hydroxyapatite, respectively.

NEW FILTRATION MODEL OF FLOCCULATED SUSPENSION

A phase transition from a dispersed suspension to a flocculated suspension occurs at a critical applied pressure (ΔP_{tc}) for nanoparticles of 20–800 nm size[11]. As a result, the consolidation behavior of the flocculated particles deviated from the filtration theory for well dispersed particles. Figure 8 shows the schematic structure of the suspension and the hydraulic pressure profile across the mold, consolidated layer and flocculated suspension. In our previous papers[11,19], we developed a new filtration theory for a flocculated suspension. The relation between suspension height and filtration time at a constant pressure, ΔP_s (= P_t–P_i), for an initially flocculated suspension is calculated by Eq. (9),

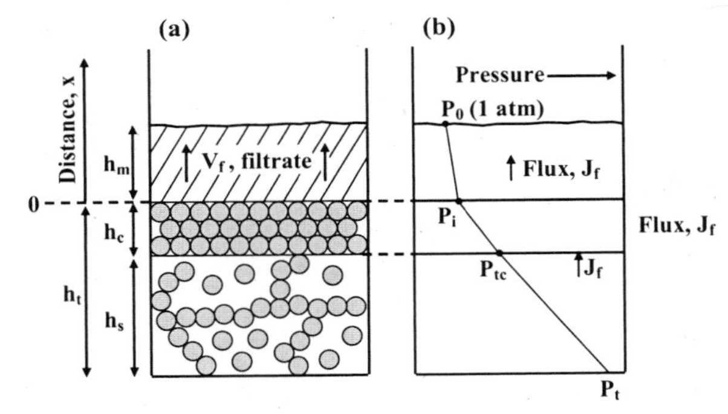

Figure 8. Schematic structure of colloidal system (a) after the phase transition from dispersed to flocculated suspension and hydraulic pressure profile of the filtration system (b).

$$\Delta P_s t = \eta BS^2 (H_0 C_0)^2 \int_{H_0}^{h_s} \frac{h_s^2}{(h_s - H_0 C_0)^3} dh_s = \eta BS^2 (H_0 C_0)^2 \left\{ \frac{1}{2} \left[\frac{h_s^2}{(h_s - H_0 C_0)^2} - \frac{H_0^2}{(H_0 - H_0 C_0)^2} \right] + \left[\frac{h_s}{(h_s - H_0 C_0)} - \frac{H_0}{(H_0 - H_0 C_0)} \right] + ln\left(\frac{H_0 - H_0 C_0}{h_s - H_0 C_0} \right) \right\} \tag{9}$$

where η is the viscosity of filtrate depending on temperature , BS^2 the constant value, B the ratio of shape factor to the tortuosity constant, S the ratio of the total solid surface area to the apparent volume, H_0 the initial suspension height, C_0 the initial volume fraction of colloidal particles, t the filtration time. The electrolyte condition (concentration, pH, other ions) affects S value[19]. This equation is completely different from Eq. (10) for the consolidation of well dispersed particles,

$$\Delta P_t t = \frac{\eta \alpha_c}{2n} (H_0 - h_t)^2 \tag{10}$$

where ΔP_t is the total pressure drop ($P_t - P_0 = \Delta P_t$), α_c the specific resistance of porous consolidated layer, h_t the height of piston, n the system parameter ($\equiv (1 - C_0 - \varepsilon_c)/C_0$, ε_c is the volume fraction of voids in the consolidated layer and affected by the electrolyte condition).

Figure 9 shows the relationship between filtration time and h_t (a) and specific resistance of filtration (b) for 5 vol% suspension (pH 7) of 30 nm SiC with polyacrylic ammonium (3.8 mass%) at 10 MPa[20]. The measured height of piston was fitted by Eqs. (9) and (10). The established filtration theory (Eq. (10)) deviates from the experimental result. However, a good agreement was found between Eq. (9) for a new filtration theory and the measured result. The specific resistance of filtration experiment (α_s) was determined by Eq. (11)[19, 20],

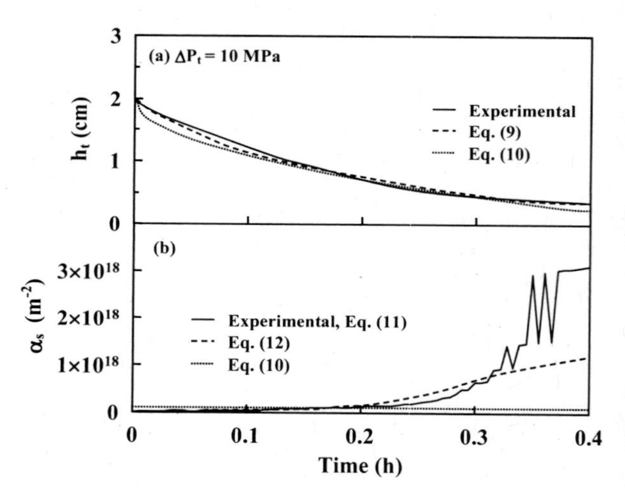

Figure 9. Relationship between filtration time and (a) height of piston and (b) specific resistance of filtration for the suspension of 30 nm SiC with PAA (3.8 mass%).

$$\alpha_s\ (observed) = \frac{(\Delta P_t / h_t)}{\eta(-\Delta h_t / \Delta t)} \tag{11}$$

The Δt was set to 1 minute to measure the difference of h_t. The observed α_s increased gradually with increasing filtration time. The relatively large difference of α_s values was measured between the experiment and calculation by Eq. (10). In Eq. (10), the α_s is treated as a constant value. On the other hand, Eq. (12) of the specific resistance for a flocculated suspension can represent time dependence of α_s using the measured h_t value[19, 20]. This model provides a good agreement of α_s values between the experiment and calculation.

$$\alpha_s = BS^2 (H_0 C_0)^2 \frac{h_s}{(h_s - H_0 C_0)^3} \tag{12}$$

FURTHER IMPROVEMENT OF FILTRATION MODEL OF FLOCCULATED SUPENSION

In previous section, the S value in Eqs. (9) and (12) was treated as a constant value. For spherical particles of diameter D, S is related to D and ε (the volume fraction of voids in the consolidated layer) by Eq. (13).

$$S = \frac{6(1-\varepsilon)}{D} \tag{13}$$

In the filtration of a flocculated suspension, (1−ε) value is represented by Eq. (14).

$$1-\varepsilon = C_0 \frac{H_0}{h_s} \tag{14}$$

The combination of Eqs. (13) and (14) provides Eq. (15) as a function of suspension height.

$$S = (\frac{6C_0 H_0}{D}) \frac{1}{h_s} \tag{15}$$

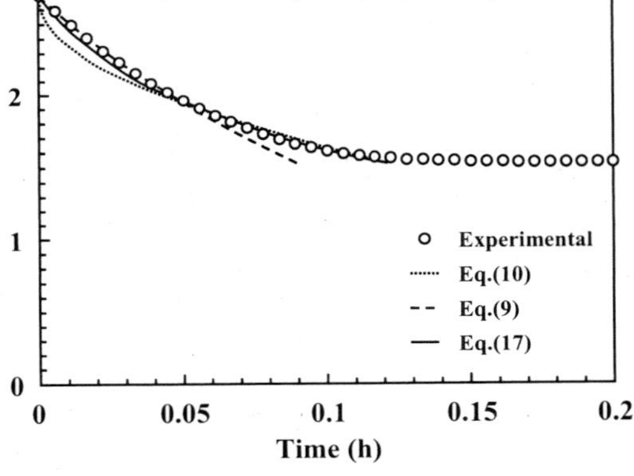

Figure 10. Suspension height of 5 vol% SiC suspension (30 nm diameter, pH 7) at 1 MPa.

Eq. (15) is substituted for Eq. (12) to give Eq. (16).

$$\alpha_s = (\frac{36B}{D^2})(H_0 C_0)^4 \frac{1}{h_s (h_s - H_0 C_0)^3} \tag{16}$$

The filtration process under a constant pressure is represented by Eq. (17) using Eq. (16).

$$\Delta P_s t = \int_{H_0}^{h_s} \eta \alpha_s h_s dh_s = \eta \left(\frac{36}{D^2}\right) (H_0 C_0)^4 \int_{H_0}^{h_s} \frac{1}{(h_s - H_0 C_0)^3} dh_s$$
$$= \left(\frac{18B}{D^2}\right) \eta (H_0 C_0)^4 \left[\frac{1}{(H_0 - H_0 C_0)^2} - \frac{1}{(h_s - H_0 C_0)^2}\right] \tag{17}$$

On the other hand, the filtration process at a constant crosshead speed (v) is represented by Eq. (18).

$$\Delta P_s = \eta v \int_{H_0}^{h_s} \alpha_s dh_s = \eta v \left(\frac{36B}{D^2}\right)(H_0C_0)^4 \int_{H_0}^{h_s} \frac{1}{h_s(h_s - H_0C_0)^3} dh_s$$
$$= \eta v \left(\frac{36B}{D^2}\right)(H_0C_0)^4 \left\{ -\frac{1}{2H_0C_0}\left[\frac{1}{(h_s - H_0C_0)^2} - \frac{1}{(H_0 - H_0C_0)^2}\right] \right.$$
$$\left. + \frac{1}{(H_0C_0)^2}\left[\frac{1}{(h_s - H_0C_0)} - \frac{1}{(H_0 - H_0C_0)}\right] + \frac{1}{(H_0C_0)^3}\left[ln\left(\frac{h_s - H_0C_0}{H_0 - H_0C_0}\right) - ln\left(\frac{h_s}{H_0}\right)\right]\right\} \quad (18)$$

Figure 10 shows the suspension height of 5 vol% SiC suspension (pH 7) of 30 nm at 1 MPa as a function of filtration time. The result is well represented by Eqs. (9) and (17) rather than established Eq. (10). In the early stage, both the equations of (9) and (17) express well the experimental result. However, the result at a longer filtration time is well represented by Eq. (17) rather than Eq. (9). This fact indicates that the treatment of S as a function of h_s (Eq. (15)) gives a better agreement between the experiment and theory (Eq. (17)). Similarly the pressure drop during the filtration of 7 vol% alumina suspension (150 nm diameter) at a constant crosshead speed (v = 0.2 mm/min) was compared with Eq. (18) and Eq. (19) for a constant S in Fig. 11.

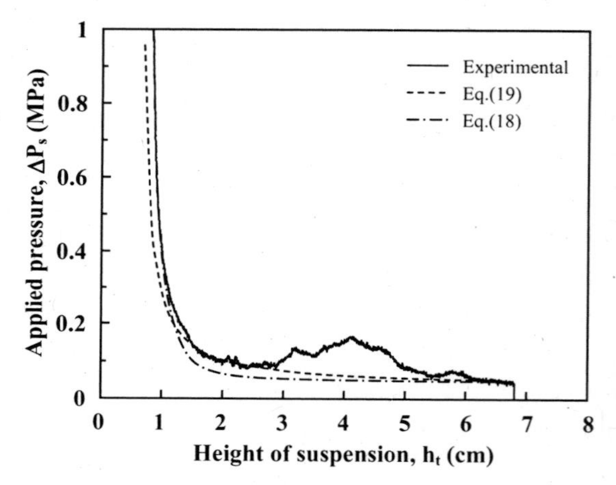

Figure 11. Comparison of ΔP_t - h_t relation between experiment and proposed theories (Eqs. 18, 19) for the pressure filtration process of 7 vol% alumina suspension at pH 7.0.

$$\Delta P_s = \eta v B S^2 (H_0C_0)^2 \frac{1}{2}\left\{\frac{H_0}{H_0^2(1-C_0)^2} - \frac{h_s}{(h_s - H_0C_0)^2} + \frac{1}{H_0(1-C_0)} - \frac{1}{(h_s - H_0C_0)}\right\} \quad (19)$$

As seen in Fig. 11, the experimental result was well represented by Eqs. (18) and (19). In the final region of filtration, Eq. (18) was in agreement with the experimental result.

CONCLUSIONS

The combination of thermodynamics of colloidal suspensions and DLVO theory succeeded in constructing colloidal phase diagrams in a map of surface potential (ϕ) and solid content of particles of 10–1000 nm diameters. The surface potential for the solidus line ($\phi = \phi_0$) reflects the enthalpy (ΔH_{ml}) for the phase transition from dispersed to flocculated suspension. The solid composition of liquidus line is related to the packing density after consolidation. The experimentally determined packing density agreed with the prediction from the phase diagram. The consolidation behavior of nanometer-sized particles at 20–800 nm was examined using a pressure filtration apparatus. The phase transition from a dispersed suspension to a flocculated suspension occurs when applied pressure exceeds a critical pressure (ΔP_{tc}). The factors affecting the phase transition of colloidal suspension are zeta potential, concentration of suspension, size of particles, polyelectrolyte dispersant and electric field. Based on the colloidal phase transition, new filtration theories were developed for a flocculated suspension. Experimentally measured piston height, specific resistance of filtration at a constant pressure or applied pressure at a constant crosshead speed of piston are well explained by the developed theories.

REFERENCES

[1]F. F. Lange, B. I. Davis and E. Wright, Processing-Related Fracture Origins: IV, Elimination of Voids Produced by Organic Inclusions, *J. Am. Ceram. Soc.*, **69**, 66–99 (1986).
[2]I. A. Aksay, F. F. Lange and B. I. Davis, Uniformity of Al_2O_3-ZrO_2 Composites by Colloidal Filtration, *J. Am. Ceram. Soc.*, **66**, C190–C190 (1983).
[3]Y. Hirata, A. Nishimoto and Y. Ishihara, Effects of Addition Polyacrylic Ammonium on Colloidal Processing of Alpha–Alumina, *J. Ceram. Soc. Japan*, **100**, 983–990 (1992).
[4]J. A. Lewis, Colloidal Processing of Ceramics, *J. Am. Ceram. Soc.*, **83**, 2341–2359 (2000).
[5]Y. Hirata, Theoretical Aspects of Colloidal Processing, *Ceram. Inter.*, **23**, 93–98 (1997).
[6]Y. Hirata and Y. Tanaka, "Characterization of Colloidal Suspension of Ceramic Nanoparticles", *Bull. Ceram. Soc. Japan*, **42**, 87–92 (2007) (in Japanese).
[7]Y. Hirata, S. Nakagama and Y. Ishihara, Calculation of Interaction Energy and Phase-diagram for Colloidal Systems, *J. Ceram. Soc. Japan*, **98**, 316–321 (1990).
[8]Y. Hirata, H. Haraguchi and Y. Ishihara, Particle-size Effects on Colloidal Processing of Oxide Powders, *J. Mater. Res.*, **7**, 2572–2578 (1992).
[9]Y. Hirata and Y. Tanaka, Thermodynamics of Colloidal Suspensions, *J. Ceram. Process. Res.*, **9**, 362–371 (2008).
[10]I. A. Aksay, C. H. Schilling, pp.85–93 in Advanced in Ceramics, Vol. 9, Forming of Ceramics, Edited by J. A. Mangels and G. L. Messing, The American Ceramic Society, Columbus, Ohio (1984).
[11]Y. Hirata and Y. Tanaka, Pressure Filtration Model of Ceramic Nanoparticles, *J. Am. Ceram. Soc.*, **91**, 819–824 (2008).
[12]J. T. G. Overbeek, in "Colloid Science I" (Elsevier Publishing Company, Amsterdam, 1952, Edited by H. R. Kruyt) pp.245–301.
[13]D. J. Shaw, "Introduction to Colloid and Surface Chemistry" (Butterworth, London, 1980), pp.183–212.
[14]W. D. Kingery, H. K. Bowen and D. R. Uhlmann, "Introduction to Ceramics, Second Edition" (John Wiley & Sons, New York, 1976) pp.227–232.
[15]H. A. Barness, J. F. Hutton and K. Walters, "An Introduction to Rheology" (Elsevier Science Publishers, Amsterdam, 1989) p.118.
[16]Y. Fukuda, T. Togashi, M. Naitou and H. Kamiya, Analysis of Electrosteric Interaction of Polymer Dispersant in Dense Alumina Suspensions with Different Counter-Ion Densities using

an Atomic Force Microscope, *J. Ceram. Soc. Japan*, **109**, 516–520 (2001).
[17]Y. Hirata, M. Nakamura, M. Miyamoto, Y. Tanaka and X. H. Wang, Colloidal Consolidation of Ceramic Nanoparticles by Pressure Filtration, *J. Am. Ceram. Soc.*, **89**, 1883–1889 (2006).
[18]Y. Hirata, H. Uchima, Y. Tanaka and N. Matsunaga, The Effect of Electric Field on Pressure Filtration of Ceramic Suspensions, *J. Am. Ceram. Soc.*, **92**, S57–S62 (2009).
[19]Y. Tanaka, Y. Hirata, N. Matsunaga, M. Nakamura, S. Sameshima and T. Yoshidome, Pressure Filtration of Nanometer-Sized SiC Powder, *J. Ceram. Soc. Japan*, **115**, 786–791 (2007).
[20]Y. Hirata, Y. Tanaka, S. Nakagawa and N. Matsunaga, "Pressure Filtration of Colloidal SiC Particles", *J. Ceram. Process. Res.*, **10**, 311-318 (2009).

THIN FILM NANOCOMPOSITES FOR THERMOELECTRIC APPLICATIONS

Otto J. Gregory, Ximing Chen, Matin Amani, Brian Monteiro and Andrew Carracia
Department of Chemical Engineering, University of Rhode Island
Kingston, RI 02881, USA

ABSTRACT

Thin film nanocomposites comprised of refractory metals and alumina were initially considered for thermal barrier coatings, since the large number of interfaces between the ceramic and metallic phases lead to considerable phonon scattering and ultra low thermal conductivity. When nanocomposites based on NiCoCrAlY and alumina were optimized for electrical and thermal conductivity, they showed considerable promise as thermoelectrics. Therefore, by replacing the alumina phase in these nanpocomposites with wide bandgap oxide semiconductors such as indium tin oxide (ITO) and zinc oxide (AZO), nanocomposites with extremely large and repeatable Seebeck coefficients were realized. Sputtering was selected as the deposition method of choice to prepare the nanocomposites, since it is a non-equilibrium, low deposition temperature process and thus, greatly reduces the tendency for particle agglomeration during deposition. Composite sputtering targets were prepared by plasma spraying optimized mixtures of NiCoCrAlY and ITO onto a stainless steel backing plates. The resulting NiCoCrAlY:ITO nanocomposite films exhibited thermoelectric powers on the order of 8000μV/°C at 1125°C, making them suitable for high temperature energy harvesting devices.

INTRODUCTION

The development of propulsion systems employing advanced materials and designs requires the continuous, in-situ monitoring of engine components operating under extreme conditions. Sensors capable of providing reliable data within the harsh environment of the gas turbine engine are needed for verification of structural models and propulsion health monitoring. Thin film semiconductive oxides and nanocomposites based on semiconductive oxides are being developed as sensors for these harsh environments, specifically thermocouples and thermoelectric devices[1-4]. Since large temperature gradients exist between the tip and the root of the turbine blade, these nanocomposite thin film thermoelectric devices should be able to generate enough electricity to power active wireless sensors.

Al_2O_3/NiCoCrAlY nanocomposites were originally studied as a candidate material for thermal barrier coating (TBC) and ceramic thermocouples, because of their ultra low thermal conductivity due to the large number of interfaces in the direction of heat transfer and associated phonon scattering[5,6]. In order to improve the thermoelectric properties, the alumina phase was replaced with an oxide semiconductor phase that exhibits a high Seebeck coefficient, i.e. ITO having a nominal composition of 90% In_2O_3 and 10% SnO_2. In doing so, thin films with both low thermal conductivity and high electrical conductivity were investigated, with the goal to produce an ideal material for high output thermoelectric devices. Using this approach, thermocouple junctions of the most responsive "n" type and "p" type ITO/NiCoCrAlY nanocomposites were produced. These devices are capable of generating thermoelectric voltages on the order of three volts, making them well suited for energy harvesting applications, and powering wireless transmitters on high temperature components.

EXPERIMENTS

To investigate the thermoelectric properties of the bi-ceramic junctions, thin films were deposited on 190mm x 25mm high purity (98.6%) alumina substrates (CoorsTek Inc.). All films were sputtered in either an MRC 822 or an MRC 8667 r.f. sputtering system. Prior to deposition of the ceramic films, the substrates were cleaned with methanol, ethanol and deionized water. Platinum films (99.99%) were used as the reference electrodes and as bond pads to make connection to ceramic elements. Prior to sputtering, the thermocouple legs and bond pads were patterned using DuPont's MX5050 dry film negative photoresist in conjunction with a liftoff technique, which permitted sputtering of films as thick as 50μm, due to its excellent resistance in plasma environments. The exposure was done using an Optical Associate's LS30 collimated UV light source with a contact photomask. The bond pads and reference electrodes were formed using a 0.75μm platinum (99.99%) film deposited from a 100mm diameter target. The ITO-NiCoCrAlY nanocomposite combinatorial libraries were deposited by co-sputtering from 150mm diameter ITO and NiCoCrAlY targets. Here, the intent was to determine the composition of the nanocomposite library with largest thermoelectric response. The optimized composition then was used to thermally spray a composite target to obtain large area coverage of the desired material.

Both platinum and nanocomposite thermoelements were sputtered at an r.f. power of 350W in pure argon ambient with total pressure of 1.2Pa (9mtorr). All ITO films were prepared by rf sputtering using an rf power of 350W (power density of $1.92 W/cm^2$) and 1800V. A background pressure $<1x10^{-4}$Pa was maintained in the vacuum chamber prior to sputtering and a high-density ceramic target with a nominal composition of 90wt% In_2O_3 and 10wt% SnO_2 was used for all ITO depositions. To evaluate the thermoelectric properties of the ITO elements prepared under different sputtering conditions, the oxygen and nitrogen partial pressures were systematically varied from zero to 0.4Pa (3mtorr) while the argon partial pressure was held constant at 1.2Pa. The thickness of the deposited thin film was measured using a DEKTAK II surface profilemeter. The nominal thickness of the as-deposited nanocomposite films was in the range 2 to 4μm. With such a small variance it is likely that there is not a thickness affect.

Figure 1 shows a schematic of the test bed used to evaluate the thermoelectric response and the location of the Type S and Type K thermocouples affixed to the ends of the alumina substrate to measure the hot and cold junction temperature, respectively. The alumina substrate was placed into the 175mm hot zone of a Deltech tube furnace where a temperature gradient was applied along the length of the sample, by placing a heat shield in the middle of the plate. The cold junction was maintained at room temperature using a water-cooled aluminum block. The furnace was thermally cycled from 200 to 1200°C at a heating/cooling rate of 3°C/min and all experiments were done in air. The hot and cold junction temperatures as well as the emf generated were monitored using a USB Data Acquisition system (PersonalDAQ 54 by I/O Tech) with "Personal DaqView" software.

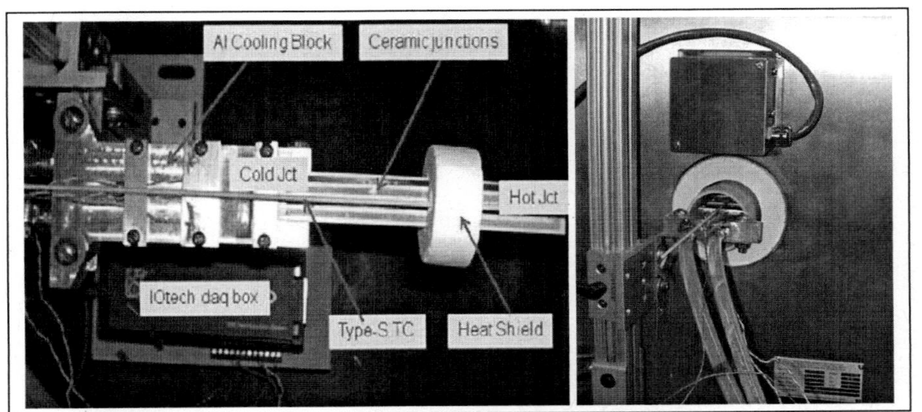

Figure 1. Photograph of the test bed used for high temperature thermoelectric measurements. Photograph on the left shows the test bed prior to being loaded into the furnace and the photograph on the right shows the test bed after being loaded into the furnace.

RESULTS AND DISCUSSION

Combinatorial materials synthesis was used to develop a new class of semiconductive oxide/refractory metal nanocomposite thermoelements. Multiple combinatorial libraries (24 thermocouple libraries per plate), each consisting of a dissimilar metal join to a platinum reference electrode, were fabricated by co-sputtering from ITO and NiCoCrAlY targets respectively (Figure 2). The left hand side of Figure 2 for example represents nanocomposite libraries with relatively high ITO content, and the right hand side of Figure 2 represents nanocomposite libraries with a relatively high NiCoCrAlY content. The thermoelectric responses were measured by inducing a thermal gradient between the hot junction and cold junction of the thin film thermocouples. A soldering iron was placed near the hot junction of the ceramic substrate, generating a reproducible temperature gradient from the hot junction to the cold junction; i.e. the soldering iron generated a junction temperature of 200°C and the resulting thermal characteristics of the ceramic plate as well as typical thermoelectric outputs with the applied temperature gradient are shown in Figure 3 below.

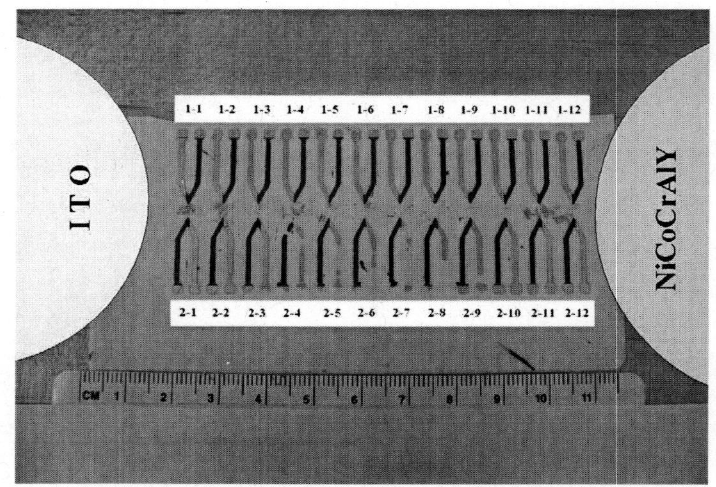

Figure 2. Layout and identification of various ITO/NicoCrAlY combinatorial libraries.

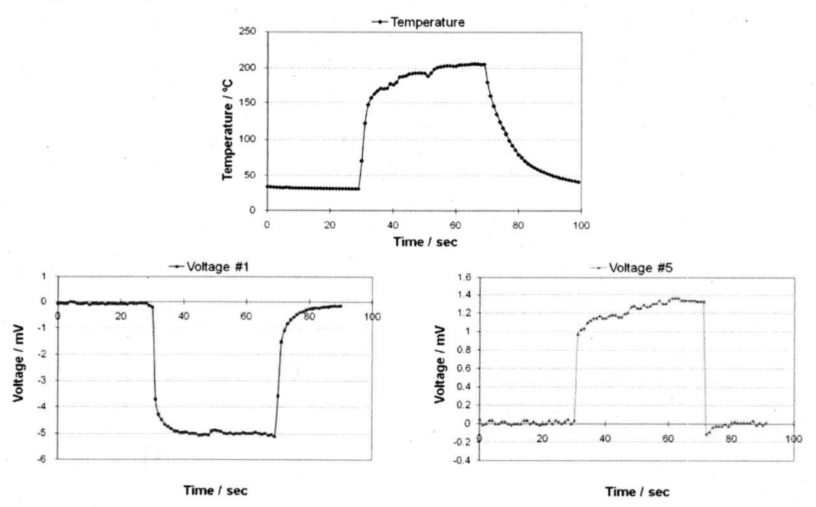

Figure 3. Temperature profile at the hot junction of each thin film combinatorial thermocouples and the corresponding thermoelectric output of combinatorial libraries #1 and #5.

The most responsive combinatorial libraries in terms of thermoelectric response were the ITO-rich libraries. Of all the ITO-rich combinatorial libraries, library # 1-1 produced the largest voltage. The NiCoCrAlY-rich combinatorial libraries had significantly lower thermoelectric responses than the ITO-rich combinatorial libraries relative to platinum. A bar graph summarizing the thermoelectric responses for the 12 most responsive libraries (ITO/NiCoCrAlY nanocomposite libraries) is shown in Figure 4. It should be noted that the "sign" of the thermoelectric response for these ITO/NiCoCrAlY nanocomposites relative to platinum changed from negative to positive depending on the relative metal content in the nanocomposite. This sign change was dependent on the position of the ceramic substrate relative to the ITO and NiCoCrAlY sputtering targets. It should also be noted that near the transition from a negative response to a positive response there were several anomalies associated with libraries #3 and #7. At this time, there is no explanation for the observed responses of these two libraries. Consistent with the change in sign of the thermoelectric response was the change in the type of the semiconductor; i.e. the change from "n" type to "p" type behavior. Since ITO is an "n" type semiconductor and NiCoCrAlY is a "p" type dopant, depending on their relative amounts in the nanocomposite, the semiconductor "type" changed from "n" type to "p" type. A hot probe technique was used for determining the type of semiconductor and this was verified by the "type" change of the semiconductor nanocomposites.

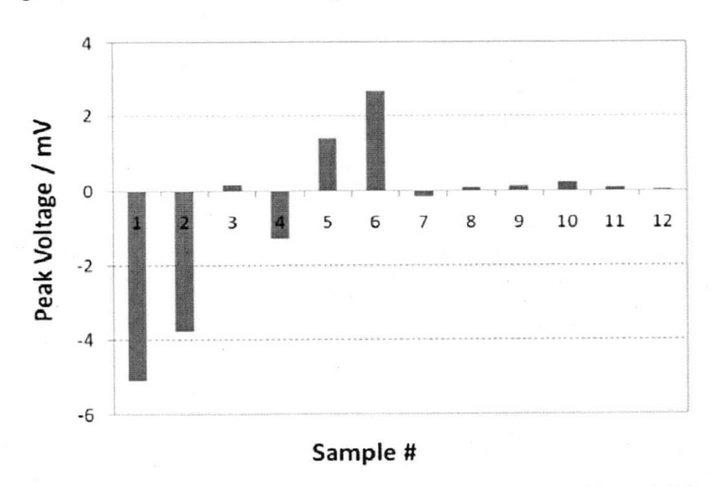

Figure 4. Thermoelectric responses of various ITO-NiCoCrAlY combinatorial libraries vs platinum reference electrodes (peak voltage at 200°C)

Based on these results, full length thermocouples were fabricated by sputtering from plasma sprayed targets using the same sputtering conditions as those used to prepare the combinatorial libraries. Since it is well documented that thermocouples with the largest Seebeck coefficients are obtained when "n" type material is combined with "p" type material[7,8], it was not surprising that the most responsive combinations overall were produced from nanocomposite thermoelements high in ITO content (n-type) and low in NiCoCrAlY content. The "p" type ITO/NiCoCrAlY nanocomposite was combined with platinum, ITO and "n" type ITO/NiCoCrAlY to form thermocouples and their thermoelectric response after repeated thermal cycling is shown in Figures 5, 6 and 7, respectively. The thermoelectric response of the Pt vs ITO/NiCoCrAlY nanocomposite thermocouple is one order of magnitude smaller than those of bi-ceramic junctions. This is because platinum has a much larger thermal conductivity than either of the nanocomposite or ceramic legs. However its Seebeck coefficient is quite large (1200μV/°C) compared to a Pt vs ITO thermocouple (78μV/°C) which is more typical of a metal vs "n" type semiconductor thermocouple. From here one can see the advantage of "p" type semiconductor vs "n" type semiconductor for energy harvesting. The ITO vs "p" type ITO/NiCoCrAlY and "n" type ITO/NiCoCrAlY vs "p" type ITO/NiCoCrAlY produced maximum thermoelectric outputs on the order of 3000mV and 2100mV, respectively. These devices survived tens of hours of exposure at 1160°C where they exhibited a stable and repeatable response. Note that the thermoelectric voltage of the platinum vs ITO/NiCoCralY thermocouple increased steadily during the dwell at 1230°C and is attributed to annealing effects in the thin films, since the films were only exposed to temperatures on the order of 800°C during the initial annealing in nitrogen. In Figures 6 and 7 a sharp drop in thermoelectric voltage was observed at a hot junction temperature of 1200°C. The exact cause for this drop in voltage has not been determined to date but the most likely cause for this behavior is the depletion of excess charge carriers by a compensation mechanism. However, the sharp drop or spike in thermoelectric output disappeared when the hot junction temperature was cycled one more time from 1200°C to room temperature (Figure 6).

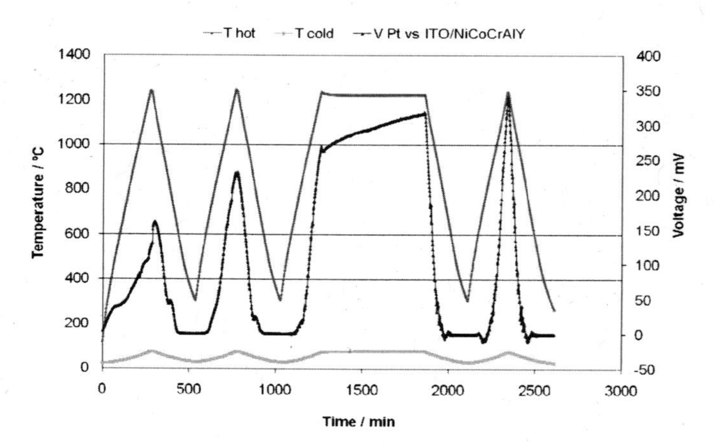

Figure 5. Thermoelectric response of a Pt vs "p" type ITO/NiCoCrAlY nanocomposite (co-sputtered) junction.

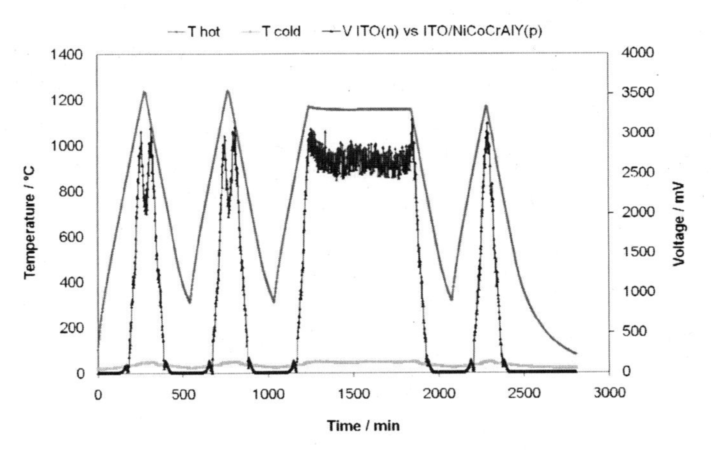

Figure 6. Thermoelectric response of an "n" type ITO nanocomposite vs "p" type ITO/NiCoCrAlY nanocomposite (co-sputtered) junction.

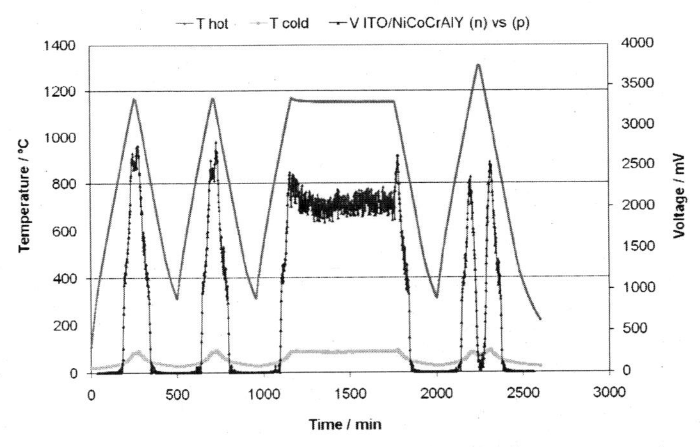

Figure 7. Thermoelectric response of an "n" type ITO/NiCoCrAlY nanocomposite vs "p" type ITO/NiCoCrAlY nanocomposite (co-sputtered) junction.

"P" type ITO/NiCoCrAlY nanocomposites were also prepared by sputtering from thermally sprayed targets having a nominal composition of 85wt% ITO and 15wt% NiCoCrAlY. When these nanocomposites were combined with platinum or "n" type ITO (including an 90wt% In_2O_3/10wt% SnO_2 and an 95wt% In_2O_3/5wt% SnO_2 target), very large thermoelectric voltages were observed at a peak temperature of 1125°C, as shown in figures 8-10. It was surprising to see that the Pt vs "p" type ITO/NiCoCrAlY nanocomposite thermocouples had a much larger thermoelectric response (about 2.5V) than that of the co-sputtered nanocomposite even tested at much lower temperature. The two different ITO vs "p" type ITO/NiCoCrAlY nanocomposite thermocouples had average thermoelectric outputs of 2.2V and 1.7V, respectively at a hot junction temperature of 1025°C.

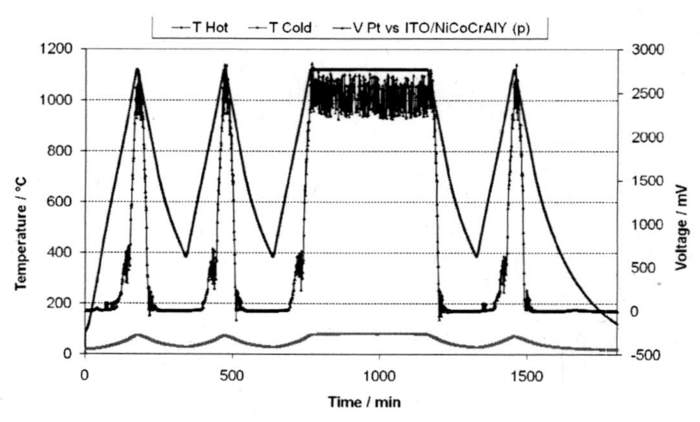

Figure 8. Thermoelectric response of a Pt vs "p" type ITO/NiCoCrAlY nanocomposite (sputtered from thermally sprayed target) junction.

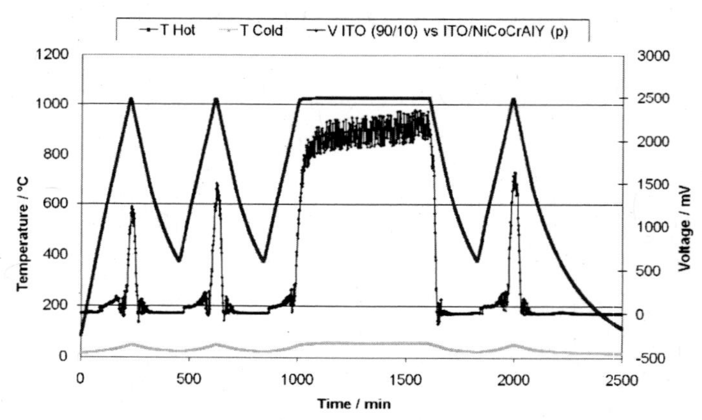

Figure 9. Thermoelectric response of an "n" type ITO (90/10) nanocomposite vs "p" type ITO/NiCoCrAlY nanocomposite (sputtered from thermally sprayed target) junction.

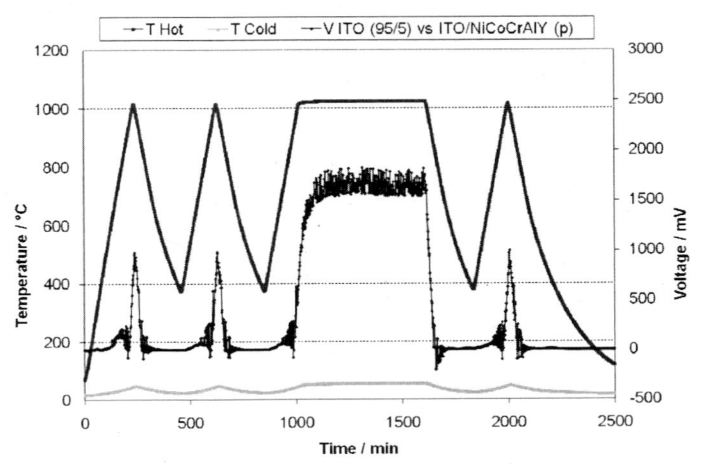

Figure 10. Thermoelectric response of an "n" type ITO (95/5) nanocomposite vs "p" type ITO/NiCoCrAlY nanocomposite (sputtered from thermally sprayed target) junction.

A summary of the thermoelectric properties of the nonacomposite bi-ceramic junctions as well as a comparison of some previous results of metal vs "n" type semiconductor thermocouples[9] are shown in Table 1. The ITO/NiCoCrAlY junctions have much larger thermoelectric output over a relatively narrow temperature range (from 700 to 1200°C). The thermoelectric output was very small when the hot junction temperature was below 700°C. To achieve large thermoelectric powers, a material which exhibits large electrical conductivity along with low thermal conductivity is desired. The reason for the small output at low temperatures was attributed to the large electrical resistivity (about 1Ωcm) of the ITO/NiCoCrAlY nanocomposite at low temperatures. By raising the temperature to 1200°C, a much higher concentration of excess charge carriers were generated, thus, reducing the resistivity to levels on the order of 0.01Ωcm. As a result the thermoelectric voltage was dramatically increased. Other bi-ceramic junctions, including ITO vs Pt and AZO vs Pt, exhibited smaller thermoelectric powers over a much wider temperature range (room temperature to 1200°C). This was attributed to the lower resistivities (0.01Ωcm) and much larger thermal conductivities relative to those of the ITO/NiCoCrAlY nanocomposite.

Table 1. Comparison of several bi-ceramic nanocomposite junctions

Junction	Max. emf mV	Seebeck Coefficient μV/°C		Hysteresis	Temperature Range °C
		Heating	Cooling		
ITO/NiCoCrAlY (p) vs Pt	350	1250	1200	small	850-1200
ITO/NiCoCrAlY (p) vs ITO (n)	3000	9300	8800	large	750-1100
ITO/NiCoCrAlY (p) vs ITO/NiCoCrAlY (n)	2100	8400	6200	large	750-1050
Al_2O_3/NiCoCrAlY vs ITO (n)*	291	700	668	large	25-1200
ITO (n) vs Pt*	77	78	78	none	25-1200
AZO (n) vs Pt	225	257	248	small	25-1200
ITO (n) vs AZO (n)	135	153	146	small	25-1200

* - *Results from Reference 9*

A mixture of 85wt% ITO and 15wt% NiCoCrAlY was thermally sprayed onto a 150mm diameter backing plate to make a nanocomposite sputtering target. The target was analyzed using a JOEL 5900 SEM in backscattered mode. The upper right SEM micrograph in figure 11 shows the surface morphology of the composite target. Note that the light phase here is ITO and the dark phase is NiCoCrAlY since the image was taken in backscatter mode. The area fraction and thus the volume fraction of the two phases is consistent with the ratio of the chemical constituents. The lower right micrograph in figure 11 is a TEM micrograph of the as-deposited film having a nominal thickness of 5nm. The TEM micrograph also shows the NiCoCrAlY matrix (light phase) uniformly distributed throughout the ITO (dark phase) and the associated electron diffraction pattern indicates that the nanocomposite is polycrystalline.

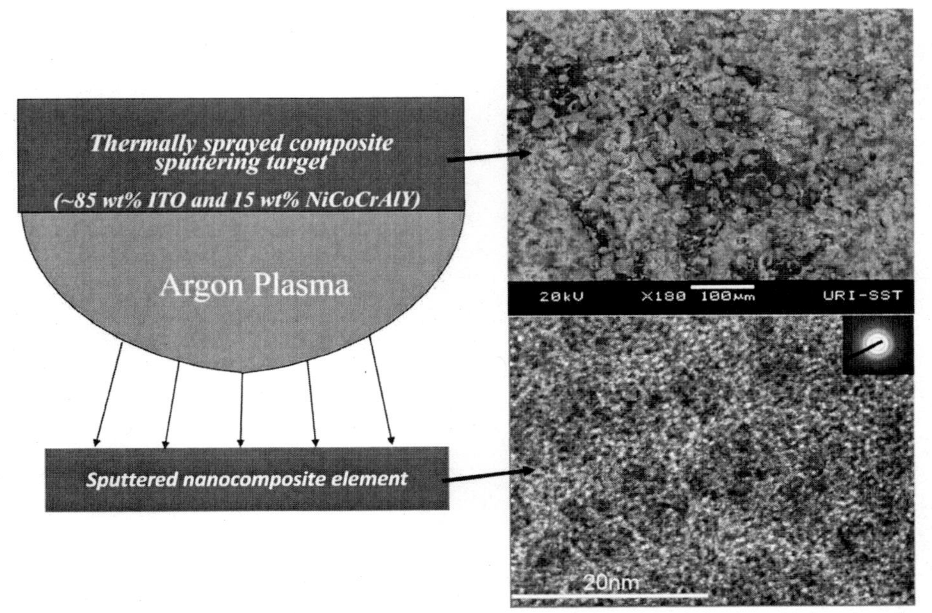

Figure 11. Schematic of the sputtering process and associated SEM micrograph showing the plasma sprayed ITO/NiCoCrAlY composite material (upper right) and resulting ITO/NiCoCrAlY nanocomposite produced by sputtering (lower right). Note: The light phase is ITO and the dark phase is NiCoCrAlY in the SEM micrograph, which is reversed in the TEM micrograph.

CONCLUSIONS

Nanocomposite thermoelement libraries comprised of ITO and NiCoCrAlY were prepared by co-sputtering from individual targets onto alumina substrates. Combinatorial chemistry was used for rapid screening purposes to determine the optimal ratio of ITO and NiCoCrAlY phases in the nanocomposite used to form these high thermoelectric output materials. Experiments showed that at relatively low NiCoCrAlY contents, a sign change in thermoelectric output was observed: i.e. a transition from an "n-type" nanocomposite to a "p-type" nanocomposite was observed. The optimized composition in terms of thermoelectric output was established from a "p" type nanocomposite containing approximately 15wt% of NiCoCrAlY. High temperature testing showed that repeatable and stable thermoelectric powers were achieved with bi-ceramic junctions prepared from a plasma sprayed composite target and an ITO target. ITO-NiCoCrAlY nanocomposites yielded the largest thermoelectric output of all nanocomposites/oxides studied to date. A maximun thermoelectric voltage on the order of 3V and a Seebeck coefficient on the order of 8000µV/°C was achieved with an "n" type ITO vs "p" type ITO/NiCoCrAlY nanocomposite thermocouple when tested in the temperature range 750-1100°C. The high threshold temperatures associated these thermocouples is due to the low concentration of excess charge carriers at low temperature. The large thermoelectric powers and large Seebeck coefficients can be attributed to the low thermal conductivity and reasonably good electrical conductivity possible from phase segregation at small length scales. Based on the large temperature gradients (approximately 450°C) that exist from the tip of the turbine blade to the root of a blade, these devices have considerable potential for energy harvesting applications.

REFERENCES

[1]O.J. Gregory and T. You, Ceramic Temperature Sensors for Gas Turbine Engine Applications, IEEE Sensors Journal, **5** (5), 833-838 (2005).

[2]O.J. Gregory, Q. Luo, and E.E. Crisman, "High Temperature Stability of Indium Tin Oxide Thin Films", Thin Solid Films, **406**, 286-293 (2002).

[3]D.M. Farrell, J. Parmar, B.J. Robbins, The Development of ceramic-based thermocouples for application in gas turbines, New Orleans, LA, USA, June 4-7, 2001, ASME TURBO EXPO (2001).

[4]O. J. Gregory and T. You, "Piezoresistive Properties of ITO Strain Sensors Prepared with Controlled Nano-Porosity", J. Electrochemical Society, **151** (8), 198-203 (2004).

[5]O.J. Gregory, M. Downey, T. Starr, S Wnuk, V Wnuk, An Intermediate TCE Nanocomposite Coating for Thermal Barrier Coatings, Materials Research Society Symposium Proceedings, **785**, 489, (2004).

[6]E. M. Meier Jackson, N. M. Yanar, M. C. Maris-Jakubowski, K. Onal-Hance, G. H. Meier, F. S. Pettit, Effect of surface preparation on the durability of NiCoCrAlY coatings for oxidation protection and bond coats for thermal barrier coatings, Materials and Corrosion, **59** (6), 494 – 500, (2008).

[7]Lon E. Bell, Cooling, Heating, Generating Power, and Recovering Waste Heat with Thermoelectric Systems, Science, **321**, 1457-1461, (2008).

[8]Yu G Gurevich, G N Logvinov and O Yu Titov, Thermoelectric Conversion in p–n Structures Based on Separation of Charge Carriers by Energy in Space Owing to Electron–Phonon Drag, Semiconductor Science Technology, **20**, 632-637, (2005).

[9]O.J. Gregory, and X. Chen, Preparation and Characterization of High Temperature Thermoelectrics Based on Metal/Oxide Nanocomposites, The Materials Research Society Symposium Proceedings, Boston, (2007).

INDIUM TIN OXIDE NANOSIZED TRANSPARENT CONDUCTIVE THIN FILMS OBTAINED BY SPUTTERING FROM LARGE SIZE PLANAR AND ROTARY TARGETS

E. Medvedovski*, C.J. Szepesi, O. Yankov
Umicore Indium Products
50 Sims Ave., Providence, RI 02909, USA

P. Lippens
Umicore Thin Film Products
p/a Kasteelstraat 7, B-2250 Olen, Belgium

ABSTRACT

Indium tin oxide (ITO) transparent conductive thin films widely used as electrode layers in optoelectronic devices, such as flat panel displays, solar cell, touch panels and some others, are usually manufactured by sputtering process, and the requirements for these films in terms of their quality and manufacturing efficiency are constantly growing. In order to obtain high quality films, high quality ITO ceramic sputtering targets and optimized sputtering process have to be used. Advanced ITO sputtering targets produced with a planar design consisting of large tiles and with a new generation rotary design based on hollow cylindrical bodies have been developed and commercialized. ITO sputtering targets with high density (up to 99.5% of TD) and uniform microcrystalline structure allow to produce nano-crystalline thin films with properties, which satisfy to industrial requirements, using optimized DC magnetron sputtering process. The obtained nanosized thin films have high electrical conductivity and transmittance (up to 93%). The influence of target configuration and the features of the sputtering process, which define thin film morphology and properties, are discussed.

1. INTRODUCTION

Highly transparent and electrically conductive oxide (TCO) thin films are widely used as electrode layers in optoelectronic devices, such as in flat panel displays (FPD), e.g. liquid crystal displays (LCD), organic light-emitting diodes (OLED), plasma display panels (PDP), touch panels, electrochromic devices, as well as antistatic conductive films and low emission coatings[1-5]. Also TCO films have had a great interest in photovoltaic applications for the formation of flexible thin film solar cells. Along with optical properties, TCO films are used for high temperature strain gas sensors. The films are commonly produced by conventional direct current (DC) magnetron sputtering on glass or polymer substrates, requiring a fine-tuned deposition process and high quality sputtering targets. Ceramic sputtering targets are used as a cathode for magnetron sputtering equipment for TCO film processing. Required film properties are defined by compositional and structural features of the ceramics and by sputtering process parameters. The ceramics should be of high purity with a uniform microcrystalline structure. They should possess high electrical conductivity and high density (99+% of TD) to maximize the useful life of the targets. These characteristics should allow the formation of electrically conductive transparent films without structural defects during sputtering.

One of the most reliable and suitable materials for sputtering targets is indium-tin oxide (ITO) ceramics because these materials provide highly homogeneous nanostructured or amorphous transparent (greater than 80% of transmittance in optical range) and electrically conductive thin films with a thickness of 70 - 250 nm[3-6]. These ceramics are formed by the addition of tin oxide to indium oxide resulting in modification (distortion) of the crystalline lattice of indium oxide and an increase its electrical conductivity. ITO ceramics have a wide direct bandgap (E_g is about 3.6 eV). Introduction of Sn^{4+} in In^{3+} cation sites in the crystalline lattice is balanced by vacancies or O^{2-} sites. This results in the donation of free electrons to the lattice and provides *n*-type electrical conductivity. The content of Sn^{4+} and final density of the ceramics are among crucial factors in the extent of electrical conductivity[6-9]. Due to the presence of lattice defects, such as interstitial atoms or oxygen deficiencies, it is not straight forward to define the theoretical density of ITO ceramics. However, it is generally accepted as 7.14-7.16 g/cm^3 for the ITO 90/10 ceramic composition. This composition (i.e. with an approximate wt.-% ratio of 90/10 between In_2O_3 and SnO_2) is one of the most widely used in industry, since it provides the high quality conductive and transparent films required for the optoelectronic applications. However, some other ITO compositions with the ratios from 98/2 to 80/20, such as 98/2, 97/3, 95/5, 85/15, 80/20 and others, are used depending on the customers' requirements in similar and other optoelectronics devices.

Quality of DC magnetron sputtered thin films is generally superior when ceramic targets have higher density; higher density targets also display an enhanced deposition rate[10, 11]. Dense ceramic targets have higher resistance against sputtering erosion and nodule formation. In particular, nodules ("black spots"), which are considered as indium sub-oxide In_2O[12-14], occur during sputtering on the periphery of the erosion race track on target surfaces and tend to cause electrical arcing. The presence of nodules deteriorates the properties of the films, and they must be periodically removed during processing. The nature of nodule formation is complex, and its mechanism has not yet been completely understood[12-17]. However, based on the experimental results, B.L. Gehman et al[10] noted that high ceramic target purity is not the major property that promotes the films with lower resistivity, i.e. ultrahigh purity grade targets would be only a small advantage to the film quality. Similarly, a small influence of ITO ceramic density on electrical resistivity and transmittance is noted by R. Yoshimura et al[11] when density is above 92-95% of TD. The authors[12, 17] suggest that sputtering parameters may have a greater influence on the film quality and nodule formation when using rather high-density and uniform ceramic targets[11].

The development and characterization of ITO and some other In_2O_3-based ceramics are of interest of many ceramic manufacturers and TCO consumers. Despite numerous studies of ITO thin films, it is not enough data describing structure and properties of commercially produced ITO ceramics and thin films obtained from these ceramics. The challenges of the manufacturing of ITO sputtering targets include the necessity of the use of high quality starting materials, especially In_2O_3 powders with respect to purity, morphology and sinterability, manufacturing routes and sintering process. Particularly, the shaping of ITO ceramic target bodies should provide the absence of deformation and cracking when sub-micron materials, which are necessary for good sinterability of the ceramics, are used. In addition, ITO ceramics, in general, have low sinterability, but the use of sintering aids is strongly limited. However, using in-house prepared In_2O_3 powders, the state-of-the-art technology of ITO ceramic large-size planar and newly developed rotary targets has been developed and commercialized at Umicore Indium Products (UIP). ITO ceramic tiles with areas up to 1200-1700 cm^2 (with a variety of dimensions) for planar targets and ITO hollow cylinders with diameters of 100-200 mm and a wall thickness of 4-10 mm, which are assembled into rotary targets with a length up to 3.8 m, are currently manufactured. Properties of the commercially manufactured ITO ceramics and the nanosized films obtained from these ceramics deposited by DC magnetron sputtering are reported; the film properties are analyzed as a function of sputtering conditions.

2. SPUTTERING TARGET DESIGN AND ITS INFLUENCE ON THIN FILM PROCESSING

Due to the present need of high quality films in large area optoelectronic devices, ceramic targets are required to be as large as possible. Till recent time, only planar sputtering targets were used in industry. These targets consist of several dense monolithic tiles, which are assembled and bonded onto a metallic backing plate. Large-size tiles with areas up to 1500-1700 cm^2 are demanded to minimize the number of joints in a target surface. The manufacturing of these large-sized fully dense products is quite challenging for commercial ceramic processing. Conventional planar targets are eroded during sputtering, resulting in a specific racetrack pattern (localized plasma zone). Only 20-40% of planar targets are utilized. Moreover, the risk of formation of nodules on the surface of the targets, which finally leads to disruption of the film deposition, is relatively high[12-17]. Thus, the need for system maintenance during target lifetime, which reduces system up-time, is inevitable using a planar cathode configuration.

In order to increase both utilization of expensive ITO ceramic sputtering targets and the efficiency of the sputtering process, rotary sputtering cathodes equipped with tubular ceramic targets may be employed[18, 19]. Such rotary sputtering targets are regarded as a new generation of sputtering targets used for TCO thin film preparation. A schematic of rotary and planar targets are shown in Fig. 1. Opposed to planar targets, almost the entire surface of rotary targets becomes a working surface that results in a significant increase of target utilization (typically 70-90% depending on the magnet array or about 3 times greater than the utilization of planar targets). Erosion of the quasi entire surface area also avoids undesired particle re-deposition on the target surface, so the growth of nodules can be avoided or, at least, significantly suppressed. Moreover, because the thermal load is on the entire cylindrical target surface area with rotary magnetrons, higher power loads can be applied on rotary targets. This normally leads to higher deposition rates. All these features provide a significant reduction of the thin film production cost, and a longer service cycle may be achieved.

Recently introduced rotary sputtering targets consisting of tubular metal substrates coated with ceramic powders used for TCO thin film preparation (e.g. ITO) are produced by employing high-temperature processing, such as plasma spray coating and some others[19]. However, these designs and processing methods do not allow to obtain high density ceramic layers (limited to 90-95%) with a necessary thickness; typically, only 2-3 mm may be obtained by these technologies, while 4-10 mm after post-firing machining are typically required for planar targets in industry. Also the ceramic layers of the targets formed by these methods cannot be easily separated from a metallic substrate that creates difficulties in recycling of expensive ITO material. The use of the rotary targets with hollow cylindrical segments made by "traditional" ceramic processing bonded to a metallic backing tube can significantly reduce these difficulties. However, ITO rotary sputtering targets were not widely used in the optoelectronic industry until present time due to serious complexity in the ITO ceramic components manufacturing. Our recent studies have allowed to implement the technology of ITO "ceramic" rotary targets to the TFP industry, increasing the process efficiency.

3. EXPERIMENTAL

3.1. Starting Materials and Manufacturing

High-purity commercially produced In_2O_3 and SnO_2 powders are used as the main starting materials for production of ITO ceramics. The In_2O_3 powders are manufactured by UIP using a proprietary process from pure indium via acidic dissolution with subsequent neutralization and precipitation of $In(OH)_3$. The obtained $In(OH)_3$ is then calcined at a proper temperature. Usually hydrochloric acid is used for acidation, and the In_2O_3 powders prepared via this route are denoted as type II. Also the "advanced" starting powder preparation process that provides a much greater homogeneity of SnO_2 in In_2O_3 is used for a "special" grade of material. Processing "waste" materials and spent targets are recycled to obtain pure In. Each lot of starting In and prepared $In(OH)_3$ and In_2O_3 powders are qualified by chemical analyses and powder characterization. In_2O_3 powder, as the major ingredient, is produced of 99.99%-purity based on total metallics. This is achieved through the accuracy and multi-step process control of In_2O_3 preparation. Typical properties of the processed powders are summarized in Table 1, and their typical particle size distributions are illustrated in Fig. 2. The In_2O_3 powders have a cubic morphology, and they are generally aggregated (Fig. 3). Properties of SnO_2 powders manufactured by other suppliers are also summarized in Table 1.

The weight ratio between In_2O_3 and SnO_2 is selected to optimize electrical properties of the ceramics and electrical properties and transmittance of the films. Based on numerous studies, the ratio of 90/10 generally provides the optimal combination of low electrical resistivity and high optical transmittance of films prepared by sputtering, and this composition is mostly used in industry at the present time. Because of this, the major studies described here were conducted for ITO 90/10; however some other ITO ceramic compositions used for optoelectronic industrial applications, which are produced at UIP, were also studied.

Starting materials are mixed and milled using ball milling or attrition milling processes based on the specially designed procedures. Attrition milling allows adequate disintegration in the significantly shorter time. The prepared water-based ceramic slurries (employing specially selected dispersing and binding agents) have workable viscosities and specific gravities suitable for further ceramic shaping. The ceramic bodies are shaped using a proprietary processing to produce either flat tiles of various dimensions or hollow cylindrical bodies. After gentle drying and dry state cutting (if required), "green" ceramic bodies are fired in high-temperature electric kilns using specially designed kiln loading and firing conditions. The optimized firing profile (firing temperature below 1600°C) and firing conditions provide near-complete densification (up to 99.5% of TD). Fired ceramic bodies are cut and ground with diamond tooling in order to attain precise dimensions, flatness or roundness (depending on shape) and surface quality, which are required for the back face metallization, bonding and sputtering processes. For example, roughness of the ground ceramics (*Ra*) is attained below 1 μm. Multi-step process control during powder preparation and ceramic manufacturing ensure high-purity and high-quality ceramic targets. Ground ceramic components are ultimately bonded to a metallic backing plate or tube. Dye penetration and ultrasonic non-destructive testing conducted for bare ceramics and for bonded targets confirmed the absence of cracks, flaws and other defects, which might negatively affect sputtering behavior. The examples of large-size planar and rotary targets and their constituents manufactured at UIP are illustrated in Fig. 4.

3.2. Sputtering

ITO films were deposited from planar and rotary ITO ceramic targets, including from the targets with actual dimensions used for industrial applications, onto glass substrates (Corning No. 1737F) using industrial DC magnetron sputtering systems in collaboration with designated sputtering equipment companies. The base pressure, power density and sputtering pressure in the process module during the sputtering runs were typical for industrial applications. The base pressure in the process module was below 5×10^{-5} Pa. Typically a power density of 3.1 W/cm^2 for planar targets and 2-3 times higher for rotary targets was applied during the sputtering runs. The sputtering pressure was in the range of 0.3-0.6 Pa and could be adjusted using a mass flow controller. Specifically for rotary targets, different sputtering powers from 5 kW up to 12 kW were applied with a total energy of greater than 3,000 kW-hr, and sputtering behavior (e.g. plasma behavior, arc formation, cycling performance, film uniformity) was evaluated. The sputtering process was run in an Ar atmosphere with addition of 0-5% of O_2 reactive gas. The sputtering was conducted at ambient and elevated temperatures, i.e. on the pre-heated substrate (up to 200-235°C); IR-radiation lamps installed in the process module were applied to maintain uniform substrate heating. Post-deposition heat treatment (annealing) of the coatings produced at ambient temperature was carried out using an electric furnace in an air atmosphere. Annealing was carried out at temperature of 200°C over 60 min. The parameters of the sputtering process were optimized in order to obtain high-quality films depending on particular conditions.

3.3. Characterization

Particle size distribution of processing powders was measured using the Sedigraph and CPS Disc Centrifuge instruments; specific surface area was measured with Brunauer-Emmett-Teller (BET) method. The morphology of the powders was studied by scanning electron microscope (SEM). Phase composition and microstructure of the fired ceramics and sputtered films were studied using glancing incidence angle X-ray diffraction (XRD) as well as SEM and transmission electron microscopy (TEM) under different magnifications using "as-received" or etched samples. Thermal gravimetric analysis (TGA) was conducted in the range of 20-1400°C using a standard procedure. Oxygen content in the ITO ceramics was determined using TGA. This method is based on the calculation of the weight decrease after reduction of the pulverized sintered ITO ceramic powder since it is assumed that the weight decrease is equivalent to the oxygen content in ITO when In and Sn are fully reduced.

Density of ceramic components was measured by the water immersion method based on Archimedes law. Young's modulus was tested using the resonant frequency method in accordance with ASTM C885. Sonic velocity was determined based on the formula related Young's modulus and sonic velocity: $E = c^2d(1+p)(1-2p)/(1-p)$, where E is Young's modulus, c is sonic velocity, d is density, p is Poisson ratio also determined by the ASTM 885. Flexural strength (four-point loading) was tested in accordance with ASTM C1161. Thermal diffusivity α was measured by the laser flash technique in accordance with ASTM E1461. Specific heat (heat capacity) C_p was tested using a differential scanning calorimeter (ASTM E1269). Thermal conductivity λ values were calculated using a formula: $\lambda = \alpha C_p d$ (where d is density). Thermal properties were determined in the temperature range of 20-250°C. Coefficient of thermal expansion (CTE) was determined in the temperature range of 20-1000°C using a quartz dilatometer. Specific electrical resistivity was determined using the four-point probe measuring unit. Test ceramic samples were cut from actual tiles-targets for the dimensions required by the appropriate testing procedure.

ITO film thickness was optically determined using a reflectometer analyzer (J.Y. Horiba, Tokyo). Film uniformity was determined by comparison of the thickness measuring at different points of the film samples. Microstructure and phase composition of the films were studied using SEM and XRD techniques, respectively. Specific electrical resistivity of the films was determined using a four-point probe measuring unit (Jandel Scientific, Corte Madera, CA). Transmittance in the visible range from 400 to 800 nm wavelength was measured using a spectrophotometer (Perkin Elmer, Norwalk, CT). Film stress was determined for selected films deposited onto Si wafers of 100 mm diameter in accordance with a standard procedure for FPD applications using a cantilever technique; an α-step profilometer was used to measure the bending contour due to the stress in the thin films.

4. RESULTS AND DISCUSSION

The developed technology of ITO ceramics includes optimized colloidal processing using high-purity starting materials, advanced shaping technique and optimized firing process, which is the UIP "know-how". The conducted studies allowed to achieve the preparation of ITO slurries from sub-micron particles with approximate solid contents of up to 84 wt.-%. Various shapes of ceramic bodies with quite large dimensions may be produced. The optimized material preparation process and shaping methods allowed to achieve green densities of 65 - 70% of TD.

As noted above, In_2O_3-based ceramics generally have a low sinterability that is related to the partial dissociation and vaporization of In_2O_3 and SnO_2 at elevated temperatures in accordance with reactions[20-22]:

$$In_2O_3\ (s) \rightarrow In_2O\ (g) + O_2\ (g)\quad (<1300^\circ C)$$
$$In_2O_3\ (s) \rightarrow 2InO\ (g) + 1/2O_2\ (g)\quad (>1300^\circ C)$$
$$SnO_2\ (s) \rightarrow SnO\ (g) + 1/2O_2\ (g)\quad (>1200^\circ C)$$

Partial vaporization of the oxides at elevated temperatures is confirmed by the results of TGA (Fig. 5). Based on the conducted studies, a specially designed firing profile and firing conditions are successfully used to achieve a high sinterability of the ITO ceramics without the use of expensive hot pressing technique that also has a limitation for large size products. Density of ITO ceramic bodies (tiles and hollow cylinders) fired in industrial conditions reached up to 7.13 g/cm^3, i.e. up to 99.5% of TD (considered as 7.15-7.16 g/cm^3), with a good reproducibility. These high density values can be achieved not only for rather small tiles with areas of 500-900 cm^2, but also for the tiles with areas of up to 1200-1700 cm^2 and for hollow cylinders with a wall thickness of 10-15 mm (before grinding). Firing of cylinders has an additional difficulty related to their substantial mass and the requirement of an accurate shape with minimal ovality; however, these difficulties have been overcome.

The ITO ceramics studied in this work consist of mostly cubic In_2O_3 (bixbyite) as the major crystalline phase. However, XRD analysis indicates the presence of a secondary phase $In_4Sn_3O_{12}$ in ITO 90/10 and 80/20 ceramics. This correlates with literature data, which regard the presence of this phase in ITO ceramics with a content of 6 at.% or more of Sn[14, 23-26]. The content of this phase is not accurately determined, and it depends not only on the composition (the ceramics with higher contents of SnO_2 have higher contents of the secondary phase), but on the firing conditions. In many cases, the detection of this secondary phase can be difficult since its peaks in the difractograms are overlapped with the peaks of the major In_2O_3 phase. The formation of this secondary phase may be regarded as positive from the densification standpoint because this new phase occupies the "space" between In_2O_3 grains. Due to the formation of the $In_4Sn_3O_{12}$ phase, it is much easier to obtain density values of 7.10-7.14 g/cm^3 for the 80/20 composition, although TD of this composition is lower. Crystallization of the SnO_2 phase was not detected that is in agreement with other studies[15, 21, 23, 24].

Microstructure of the ITO ceramics made by UIP is dense and uniform, and it consists of grains with sizes from 2 to 20 μm (Fig. 6). The grains consist of crystallites with sizes of around 50-100 nm (which vary depending on starting material preparation conditions) as determined by

XRD analysis using the Scherrer formula. The secondary phase formation inhibits grain growth that may be confirmed by the comparison of grain sizes of the studied materials. As can be seen from the SEM images, the grain sizes of ITO 95/5 ceramics, which does not contain the secondary phase, are 10-20 μm (Fig. 6a), while ITO 90/10 (Fig. 6b) and 80/20 (Fig. 6c) ceramics containing this phase have smaller grain sizes (5-10 μm and 2-7 μm, respectively). The grains cleavage can be seen in the SEM images of fracture surfaces. This provides evidence of a high extent of ceramic densification since the fracture occurs through the grains. Irregular small intergranular pores are uniformly distributed, but they are not interconnected. It may be noted that, in the case of manufacturing of large-size products, the firing profile is usually extended in accordance with general ceramic principles that results in more intensive grain growth.

Appropriate oxygen content in ITO ceramic targets is important to maintain a stable sputtering process and a wide "process window" required for consistency of the producing films. Depending on the process features, either fully oxidized or partially reduced ITO tiles are preferable. Theoretical oxygen content of ITO ceramics is not a reliable value because it depends on the method of calculation. This value may be 17.682% for ITO 90/10 if it is extracted from the mixture of 90 mass.% In_2O_3 and 10 mass.% SnO_2, or 17.243% if it is calculated from the composition $(In_{0.907}Sn_{0.093})_2O_3$, where some amount of Sn atoms substitute In atoms in the crystalline lattice. Due to the presence of a secondary $In_4Sn_3O_{12}$ phase, the oxygen content becomes even more uncertain. For our study, an oxygen content of 17.69% was considered as theoretical for ITO 90/10, 18.08% for ITO 80/20 and 17.48% for ITO 95/5. As noted, the oxygen content determination was conducted using a TGA analysis. The determined oxygen content of the studied ITO 90/10 ceramics is in the range of 17.55-17.90%, which is close to the theoretical value taking into account the measuring accuracy. In fact, the difference in actual and theoretical oxygen contents for the 80/20, 90/10 and 95/5 ITO compositions is higher for the compositions with lower In_2O_3 contents (for comparison, actual oxygen contents for 80/20 and 95/5 compositions are 17.94 and 17.43%, respectively). There is not a clear correlation between oxygen content values and densities of ITO ceramics when their densities are greater than 98.5-99% of TD. It is also difficult to explain the variation in the values of oxygen content when the actual oxygen contents for some samples are even greater than the theoretical value. Most likely, the answer can be found considering the grain boundaries in ITO ceramics.

Physical properties of the ITO ceramics produced by UIP are listed in Table 2. Their mechanical strength is at a moderate level; however, ITO ceramics are not intended for structural applications. Young's modulus and sonic velocity, in addition to structural properties, indicate the level of densification, especially in the presence of closed pores and macrodefects, and these data are useful for comparison of ceramics with the same composition, e.g. ITO materials. However, the determination of sonic velocity using the above mentioned formula may not be accurate enough for ITO ceramics because of a relatively wide range of Poisson ratio values.

The values of specific electrical resistivity of ITO ceramics are generally lower for the 90/10 compositions in comparison with some others. The values range from $(1.3 - 1.7)\times10^{-4}$ Ohm.cm for the compositions of 90/10 to $(2.5-2.7)\times10^{-4}$ and $(1.6-1.8)\times10^{-4}$ Ohm.cm for the 80/20 and 95/5 compositions, respectively. However, the measure of electrical resistivity is not a very reliable indicator of the ITO composition, especially if SnO_2 content is in the range of 5-12%. Densification of ITO ceramics has more influence on the electrical resistivity (ceramics of the same composition but with higher density usually demonstrate lower values of specific electrical resistivity) than their ultrahigh purity. However, a high total content of impurities (e.g. more than 1000 ppm) may result in not only the undesirable distortion of the In_2O_3 crystalline lattice and then in film transmittance reducing, but also the formation of electrically insulating layers between In_2O_3 grains. This may enhance the ceramic electrical resistivity and, consequently, the film resistivity. In this case, the consideration of impurities and their influence on electrical properties of the ITO ceramics should be "selective", e.g. the formation of alkali silicate, alkali earth silicate or some other glassy phases should be avoided. At the same time, some other impurities do not affect the electrical properties of ITO ceramics.

The values of electrical resistivity for ITO ceramics may vary in a rather wide range depending on the content of the less-conductive $In_4Sn_3O_{12}$ phase. This phase, as discussed above, is formed at Sn contents greater than 6 at.%. Its content and the solubility of Sn in In_2O_3 may depend on processing features, such as mixing-milling and, especially, firing conditions (e.g. temperature distribution in the kiln, oxygen level, etc.). Practically, oxygen level and sintering process also depend on the size of the ITO ceramic bodies, their loading in the furnace and "thermal mass" in the furnace (related to the products sizes and their amount in the furnace), which vary from firing to firing under actual manufacturing conditions. Depending on these factors, the content of the $In_4Sn_3O_{12}$ phase and electrical conductivity of the ceramics may vary. Elevated values of electrical resistivity in the 80/20 composition are explained by the crystallographic structure of the material and a greater extent of the less-conductive $In_4Sn_3O_{12}$ phase. However, the presence of this $In_4Sn_3O_{12}$ phase positively affects density values of ITO ceramics. Therefore, the influence of the phase composition and structure of ITO ceramics on their properties and quality of sputtered films is rather complex.

It is difficult to correlate the influence of microstructure (grain size and grain size distribution) on electrical properties of ITO ceramics. There is no discussion in the literature on this influence, even for small samples processed and fired under very uniform laboratory conditions. Considering real production conditions with variations in starting powders preparation and firing conditions, the structure may vary from firing to firing, especially if large-sized products are fired. The influence of microstructure on electrical properties may be difficult to observe. Only a small increase in conductivity with grain size growth may be noticed.

Thermal diffusivity and thermal conductivity of ITO ceramics have to be higher as possible to minimize thermal tensile stresses, which naturally occur in the targets during sputtering; otherwise, related defects may be a cause of non-uniform deposition of ITO films. The values of the thermal properties are well acceptable for the industrial needs, e.g. for sputtering process. The values of these properties are higher for the materials with higher contents of In_2O_3. The influence of ceramic densification on the thermal properties is similar to its influence on electrical conductivity; the higher the density, the greater the thermal diffusivity, the heat capacity and the thermal conductivity. The samples with higher electrical conductivity also demonstrate higher thermal conductivity. The values of thermal diffusivity and thermal conductivity decrease with a temperature increase. The change of thermal conductivity vs. temperature has a linear character, but the change of thermal diffusivity with temperature does not display such behavior (Fig. 7). Both electrical and thermal conductivity tend to be higher for the ITO samples with larger grain sizes, since the ceramics with larger grains contain fewer grain boundaries to act as resistance to electrical and thermal flow (with respect to electrical properties, this trend may be noted for the samples with a significant difference in the grain sizes). The values of CTE of ITO ceramics with different contents of SnO_2 are comparable that allows the use of similar compositions for metallization and bonding processes.

Thin films with a thickness of 70-150 nm obtained by conventional DC magnetron sputtering at different conditions were analyzed for their compositions, microstructure and physical properties. The results for the films obtained from 90/10 ITO ceramic targets, considered in industry as the most reliable composition, are reported. By adjusting the amount of oxygen inserted to the system, as well as other sputtering parameters, a wide process "window", a stable sputtering process and high quality films were obtained.

The obtained thin films had a uniform thickness; the film uniformity was determined optically by measuring thickness in various points over the large area of the substrate. The thickness difference between points of measuring was below 10% that is well accepted in the optoelectronic industry. Roughness of the films deposited at elevated temperature and after annealing was about 2 nm; these high film uniformity was obtained from both rotary and planar targets.

The glancing incidence angle XRD spectra of selected ITO layer structures were recorded and compared with JCPDS database #6-0416. The spectrum of an almost entirely amorphous structure (a-ITO) with a small amount of crystalline sites was inherent to ITO films deposited at room temperature ("as-deposited"), as demonstrated at Fig.8. A polycrystalline structure (p-ITO) spectrum showing Bragg lines of the cubic In_2O_3 phase with preferential orientation in the (222) direction (Fig. 8) was formed in the films produced with thermally activated processes. A rather small diffraction peak corresponding to the (440) orientation was also detected. This film structure, depending on the sputtering conditions, is in a good agreement with other studies[27-32]. At low 2θ angles (20-35°), the signals from the layers are superimposed with the X-ray amorphous spectrum of the glass substrate.

Morphology studies easily demonstrated the influence of sputtering conditions on the ITO film structures. The micrographs of a-ITO ("as-deposited") films showed a poor topographical contrast. A polycrystalline structure of cubic In_2O_3 grains was observed for the films deposited at elevated substrate temperatures. The films produced after annealing at 200°C also had very homogeneous polycrystalline structures (p-ITO) consisted of the In_2O_3 phase. These crystalline structures exhibited densely packed equiaxed (cubic) grains with average sizes of 10-25 nm (Fig. 9). Occasional micropores were observed at some triple junctions between grains. The evolution of some columnar grains with a size of 30-50 nm that form larger domains may also be seen in Fig. 9. Such patterns are typical for heat treated and elevated temperature-deposited ITO films[27, 28]. The XRD and microstructure studies verified that the annealing process thermally activated a complete crystallization of the amorphous film structure and, subsequently, promoted a density increase of the ITO films.

Specific electrical resistivity of the amorphous ("as-deposited") films was as low as about 600 μOhm.cm at the oxygen flow of approximately 1.6% (Fig. 10). After annealing, specific electrical resistivity of the nanocrystalline ITO structures was lowered to 200-210 μOhm.cm that is well acceptable for various FPD applications. The transmittance of the layers at 550 nm wavelength was approximately 85% with insertion of approximately 1.8% of oxygen for "as-deposited" films, transmittance was increased to more than 90% after 1 hr annealing in air (Fig. 11). The results are in a good agreement with previous studies[28-32] when an annealing process results in a relaxation of disorder bands as well as crystallization of the amorphous network, thus increasing electrical conductivity and optical transmittance of the film. Deposition on a heated substrate provided the same level of specific electrical resistivity and optical transmittance as the annealed films due to crystallization on the hot substrate.

Selected film samples obtained from spattering using planar and rotary targets were evaluated for their stress condition. The "as-deposited" films are considered as being under compressive stress. It was assumed that compression of the "as-deposited" films stemmed mainly from the thermal expansion mismatch between the ITO layer and the Si wafer substrate[31]. After the post-deposition annealing of amorphous ITO film, the film stress converted from compression to tensile. The values of approximately -200 MPa (i.e. compressive stress) were determined for as-deposited 150-nm films, while the values of approximately +(200-250) MPa (tensile stress) were determined for thin films after annealing. The transition of compressive stress in a-ITO to tensile stress in p-ITO after annealing may be due to an enormous densification as amorphous structure transforms to the crystalline phase. The obtained values are within the specifications for the stress conditions for the FPD applications (below than +/-500 MPa compressive / tensile[31]).

It was difficult to define a strong correlation between density of ITO ceramics and macroscopic properties of ITO films. Based on the studies conducted by K. Utsumi et al[6], an increase of density of ITO ceramics from 90 to 99% of TD resulted in a slight decrease of the film resistivity due to a slight increase of the carrier concentration. This change of resistivity was noted when density of ceramic samples increased from 97 to 99% of TD; however, an influence of such density increase on the film transmittance was not found. R. Yoshimura et al[11] also did not find film properties improvement with a ceramic density increase above 92% of TD. These authors also indicated that an increase of substrate temperature to 300°C during sputtering allows the use of lower-density ITO targets. However, it should be noted that all samples used for the sputtering tests in the present work had a high level of density (99% of TD or greater) that is required by the FPD industry. This high level of density of ceramics has to be maintained in order to achieve not only high quality film properties, but also to minimize the occurrence of the defects in the films and to maximize the sputtering efficiency in industrial conditions, especially in the case of large area targets.

Extensive sputtering tests (with a total sputtering energy greater than 2500 kW-hr) of the UIP rotary sputtering targets consisting of ITO hollow cylinders with medium-size ID (135 mm) and a wall thickness of 6 mm bonded to the titanium backing tube with a total length of the ceramic component of 1200 mm demonstrated acceptable results, including adequate film properties (e.g. electrical resistivity and transparency) comparable with those obtained by sputtering of planar targets under standard conditions and which satisfy industrial requirements. Due to rotation, utilization of the ceramics was significantly higher; practically, the whole target was spent with a thickness of the remaining material even below 1.5 mm for the almost entire length. No defined erosion groove ("race track") inherent to planar targets appeared. The films with thicknesses in the range of 70-100 nm had a high uniformity with deviation less than 10% that is well accepted for optoelectronic industry and comparable with a uniformity of films prepared using metallic rotary targets. A stable behavior was observed during the cycling process when the plasma was exposed to OFF and ON modes for several hundred times in order to test real manufacturing conditions. With a designed sputtering procedure, the following achievements may be outlined:

- no abnormalities with plasma ignition at a high power level (up to 12 kW/m)
- uniform and steady plasma behavior
- stable cycling performance (e.g. stable sputtering voltage over a long period of time and low arcing)
- sputtered nanostructured film with a high uniformity
- practically, no nodule and re-deposition on the target

The use of a rotary cathode allowed to apply a higher sputtering power. This permitted to increase a deposition rate, and as a result, provided a higher efficiency of film deposition.

Properties and morphologies of ITO films obtained from planar and rotary targets are similar, and they, as demonstrated above, have a high level that satisfies well to industrial requirements. However, the valuable benefit of the use of rotary targets in industry is dealt with the significantly higher film processing efficiency, about 3 times higher target utilization, significantly lower probability of nodule formation and re-deposition during sputtering due to rotation and, therefore, with reduced film processing cost in comparison with the use of "conventional" planar targets.

5. CONCLUSIONS

The developed technology of ITO ceramics using in-house prepared starting In_2O_3 powders allows to manufacture high-quality and high-density (up to 99.5% of TD) products with different dimensions and shapes. Ceramic components with areas up to 1700 cm^2 for planar and rotary sputtering targets are commercially produced. Thanks to optimized compositions, high uniformity and densification, the manufactured ceramics possess low electrical resistivity and acceptable structural and thermal properties. As a result, nanosized ITO films with nanocrystalline or amorphous structures obtained by DC magnetron sputtering with fine-tuned process parameters reveal low specific electrical resistivity and high transmittance required for applications in industrial large area and laboratory optoelectronic manufacturing. The use of new rotary sputtering targets provides a significant increase in the ITO film processing efficiency, greater utilization of expensive ITO ceramics (up to 90%) and overall cost reduction with reduced re-deposition during sputtering.

ACKNOWLEDGEMENTS

The authors are grateful to Dr. Guido Huyberechts (Umicore RDI, Belgium) for the helpful discussions and Dr. Eddy Boydens (Umicore RDI, Belgium) for the assistance in structural analyses. Assistance of Fraunhofer Institute for Thin Films and Surface Technology, FhG-IST Braunschweig (Germany) is appreciated greatly for performance of XRD and high resolution SEM analysis of thin films.

REFERENCES

[1]H. Hosono, Recent Progress in Transparent Oxide Semiconductors: Materials and Device Application; *Thin Solid Films*, 515, 2007, p. 6000-6014

[2]G.J. Exarhos, X.-D. Zhou, Discovery-Based Design of Transparent Conducting Oxide Films (Review); *Thin Solid Films*, 515, 2007, p. 7025-7052

[3]J.L. Vossen, Transparent Conducting Films, *Phys. Thin Films*, Ed. By G. Haas, M.H. Francombe, and R.W. Hoffman (Academic, New York) **9**, 1977, p. 1-71

[4]D.S. Ginley, C. Bright, Transparent Conducting Oxides, *MRS Bulletin*, **8**, 2000, p. 15-18

[5]I. Hamberg, C.G. Granquist, Evaporated Sn-Doped In_2O_3 Films: Basic Optical Properties and Applications to Energy-Efficient Windows; *J. Appl. Phys.*, 1986, **60**, R123-160

[6]K. Utsumi, O. Matsunaga, T. Takahata, Low Resistivity ITO Film Prepared Using the Ultra High Density ITO Target, *Thin Solid Films*, 334, 1998, p. 30-34

[7]G. Frank, H. Kostlin, Electrical Properties and Defect Model of Tin-Doped Indium Oxide Layers, *Applied Physics A*, 27, 1982, p. 197-206

[8]K. Sasaki, H.P. Seifert, L.J. Gauckler, Electric Conductivity of In_2O_3 Solid Solutions with ZrO_2; *J. Electrochem. Soc.*, **141**, 10, 1994, p. 2759-2768

[9]P.A. Cox, W.R. Flavell, R.G. Egdell, Solid-State and Surface Chemistry of Sn-Doped In_2O_3 Ceramics, *J. Solid State Chem.*, **68**, 1987, p.340-350
[10]B.L. Gehman, S. Jonsson, T. Rudolph, et al., Influence of Manufacturing Process of Indium Tin Oxide Sputtering Targets on Sputtering Behavior; *Thin Solid Films*, **220**, 1992, p. 333-336
[11]R. Yoshimura, N. Ogawa, T. Iwamoto, et al., Studies on Characteristics of ITO Target Materials (2); *TOSOH SMD Technical Note* TKN 2.003A
[12]S. Ishibashi, Y. Higuchi, Y. Oka, et al., Low Resistivity Indium-Tin Oxide Transparent Conductive Films. II. Effect of Sputtering Voltage on Electrical property of Films, *J. Vac. Sci. Technol. A*, **8** (3), May/June 1990, p. 1403-1406
[13]P. Lippens, A. Segers, J. Haemers, et al, Chemical Instability of the Target Surface during DC-Magnetron Sputtering of ITO-Coatings, *Thin Solid Films*, **317**, 1998, p. 405-408
[14]T. Omata, M. Kita, H. Okada, et al., Characterization of Indium-Tin Oxide Sputtering Targets Showing Various Densities of Nodule Formation, *Thin Solid Films*, **503**, 2006, p. 22-28
[15]B.G. Lewis, R. Mohanty, D.C. Paine, Structure and Performance of ITO Sputtering Targets, pp. 432-439 in *37th Annual Technical Conference Proceedings, Society of Vacuum Coaters*, Boston, MA, 1994
[16]A.D.G. Stuwart, M.W. Thompson, Microphotography of Surfaces Eroded by Ion Bombardment, *J. Mater. Sci.*, **4**, 1969, p. 56-60
[17]K. Nakashima, Y. Kumahara, Effect of Tin Oxide Dispersion on Nodule Formation in ITO Sputtering; *Vacuum*, 66, 2002, p. 221-226
[18]W. De Bosscher, K. Dellaert, S. Luys, et al., ITO Coating of Glass for LCDs, *Information Display*, **21**, N. 5, 2005, p. 12-15
[19]W. De Bosscher, H. Delrue, J. Van Holsbeke, et al., Rotating cylindrical ITO targets for large area coating, *J. of Society of Vacuum Coaters*, 48, 2005, p. 111-115
[20]J.H.W. de Wit, The High Temperature Behavior of In_2O_3, *J. Solid State Chemistry*, **13**, 1975, p. 192-200
[21]J.H.W. de Wit, M. Laheij, P.E. Elbers, Grain Growth and Sintering of In_2O_3, *J. Solid State Chemistry*, **13**, 1975, p. 143-150
[22]R.H. Lamoreaux, D.L. Hidenbrand, L. Brewer, High-Temperature Vaporization Behavior of Oxides II. Oxides of Be, Mg, Ca, Sr, Ba, B, Al, Ga, In, Tl, Si, Ge, Sn, Pb, Zn, Cd, and Hg, *J. Phys. Chem. Ref. Data*, **16**, N. 3, 1987, p. 419-443
[23]J.L. Bates, C.W. Griffin, D.D. Marchant, et al., Electrical Conductivity, Seebeck Coefficient and Structure of In_2O_3-SnO_2. *Amer. Ceram. Soc. Bull.*, **65**, N.4, 1986, p. 673-678
[24]N. Nadaud, N. Lequeux, M. Nanot, et al., Structural Studies of Tin-Doped Indium Oxide (ITO) and $In_4Sn_3O_{12}$, *J. of Solid State Chemistry*, 135, 1998, p. 140-148
[25]T. Vojnovich, R.J. Bratton, Impurity Effects on Sintering and Electrical Resistivity of Indium Oxide, *Amer. Ceram. Soc. Bull.*, **54**, N.2, 1975, p. 216-217
[26]W. J. Heward, D.J. Swenson, Phase Equilibria in the Pseudo-Binary In_2O_3-SnO_2 System; *J. Mater. Sci.*, 42, 2007, p. 7135-7140
[27]M. Kamei, Y. Shigesato, S. Takaki, Origin of Characteristic Grain-Subgrain Structure of Tin-Doped Indium Oxide Films, *Thin Solid Films*, **259**, N.1, 1995, p. 38-45
[28]R. Latz, K. Michael, M. Scherer, High Conducting large Area Indium Tin Oxide Electrodes for Displays Prepared by DC Magnetron Sputtering; *Jap. Journ. of Appl. Physics*, **30**, N. 2A, 1991, p. L.149-L.151

[29]P.K. Song, H. Akao, M. Kamei, et al., Preparation and Crystallization of Tin-doped and Undoped Amorphous Indium Oxide Films Deposited by Sputtering; *Jap. Journ. of Appl. Physics*, **38**, N. 9A, 1999, p. 5224-5226
[30]C. Guillen, J. Herrero, Polycrystalline Growth and Recrystallization Process in Sputtered ITO Thin Films; *Thin Solid Films*, **510**, N. 1-2, 2006, p. 260-264
[31]U. Betz, M.K. Olsson, J. Marthy, et al., Thin Films Engineering of Indium Tin Oxide: Large Area Flat Panel Displays Application, *Surface and Coating Technology*, **200**, N. 20-21, 2006, p. 5751-5759
[32]A.M. Gheidari, F. Behafarid, G. Kavei, et al., Effect of Sputtering Pressure and Annealing Temperature on the Properties of Indium Tin Oxide Thin Films; *Materials Science and Engineering* B, 136, 2007, p. 37-40

Table 1. Properties of Starting Powders for ITO Ceramics

Material	Purity, %	Particle size distribution, μm			Specific surface (BET), m^2/g
		d10	d50	d90	
$In(OH)_3$	99.99	1-3	7-12	15-20	8-12
In_2O_3, type II	99.99	0.8-0.9	3-5	7-8	0.7-1.0
In_2O_3, type IIB	99.99	0.05-0.1	0.4-1.3	1-2	13-18
SnO_2	99.9	0.1-0.2	0.3-0.9	2-3	4-9
ITO slip	-	0.1-0.2	0.4-0.6	1.5-2.5	5-7

Table 2. Some Physical Properties of Studied UIP ITO Ceramics

Property	90/10	95/5	80/20
Density, g/cm^3	7.07-7.12	7.07-7.12	7.07-7.14
Oxygen content, %	17.55-17.90	17.40-17.46	17.90-17.97
Flexural strength, MPa	150-180	-	-
Young's modulus, GPa	160-190	-	-
Poisson ratio	0.285-0.335	-	-
Sonic velocity, km/s	5500-6400	-	-
Specific electrical resistivity, Ohm.cm	$(1.3\text{-}1.6)\times10^{-4}$	$(1.6\text{-}1.8)\times10^{-4}$	$(2.5\text{-}2.7)\times10^{-4}$
CTE$\times10^6$, 1/K			
20-200°C	7.0-7.5	7.0-7.3	7.4-7.6
20-700°C	8.3-8.6	8.2-8.6	8.5-8.6
Thermal diffusivity, cm^2/s			
20°C	0.042-0.046	0.045-0.050	0.028-0.032
250°C	0.034-0.038	0.038-0.042	0.024-0.026
Heat capacity, W.s/g-K			
20°C	0.36-0.37	0.36-0.37	0.36-0.37
250°C	0.40-0.41	0.42-0.43	0.43-0.44
Thermal conductivity, W/m-K			
20°C	11-12	12.5-13	8-8.5
250°C	10-11	12-12.5	7.5-8

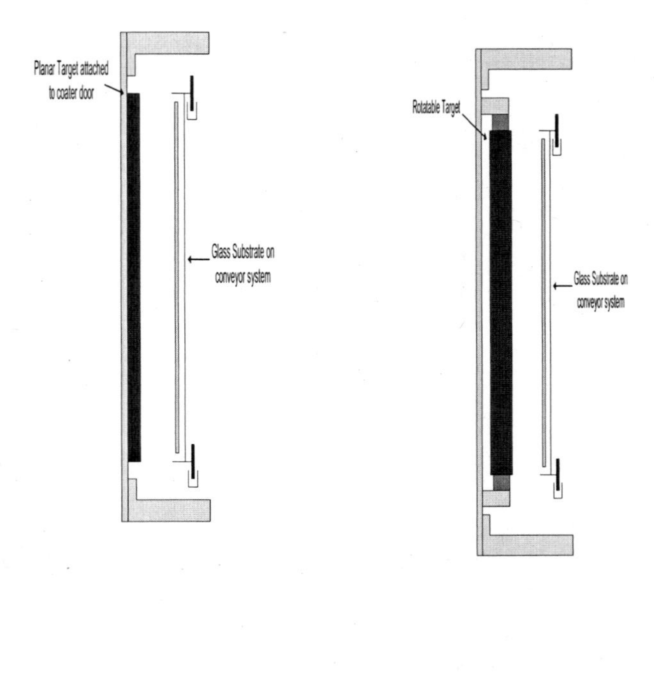

Fig. 1. Schematics of Sputtering Targets
a) Planar Sputtering Targets
b) Rotary Sputtering Targets

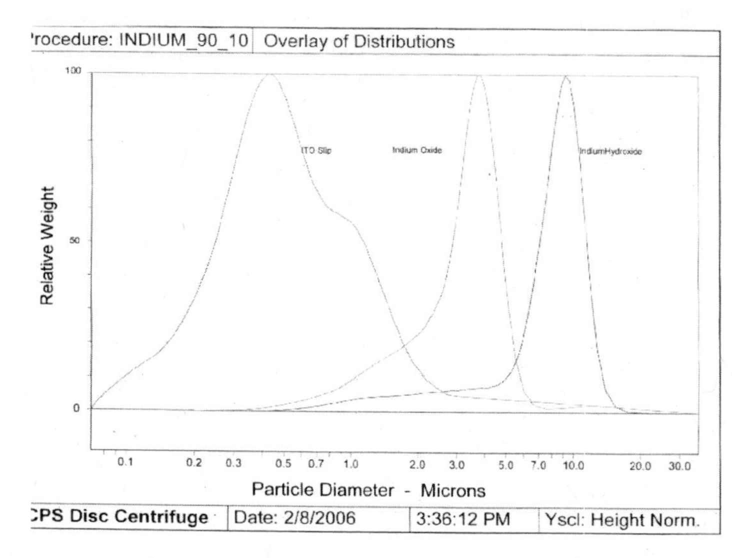

Fig. 2. Particle Size Distributions for $In(OH)_3$ and In_2O_3 Powders and ITO Slip

Fig. 3. SEM Image of In_2O_3 Type II Powder

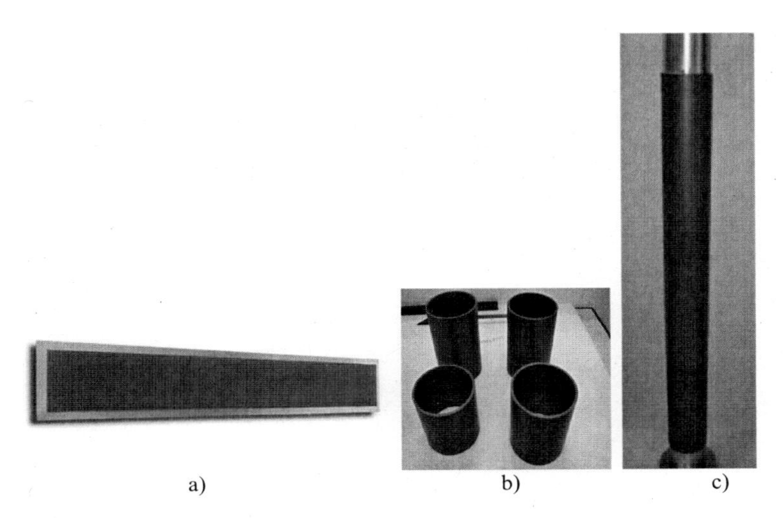

Fig. 4. ITO Ceramic Sputtering Targets Commercially Produced at UIP

a) Planar Target Consisting of 3 Large Tiles (about 1500 cm^2 each) and 2 Side Tiles
b) Ground Hollow Cylinders
c) Rotary Target with a Total Length of More than 1200 mm

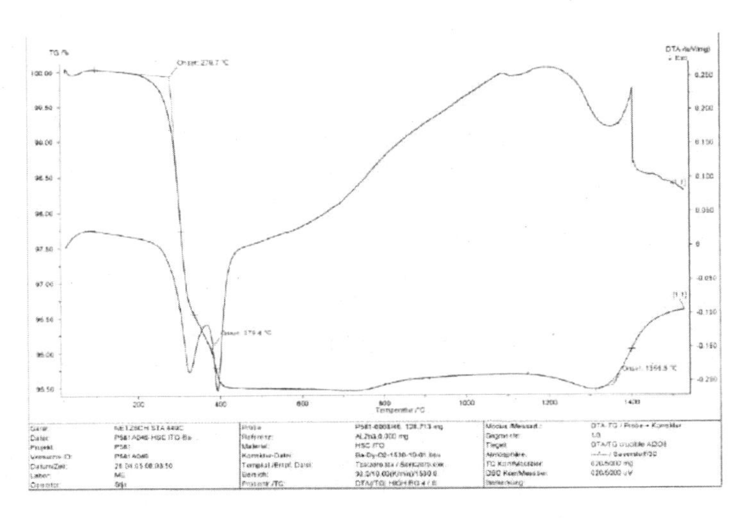

Fig. 5. TGA Curve of ITO (90/10) Ceramics

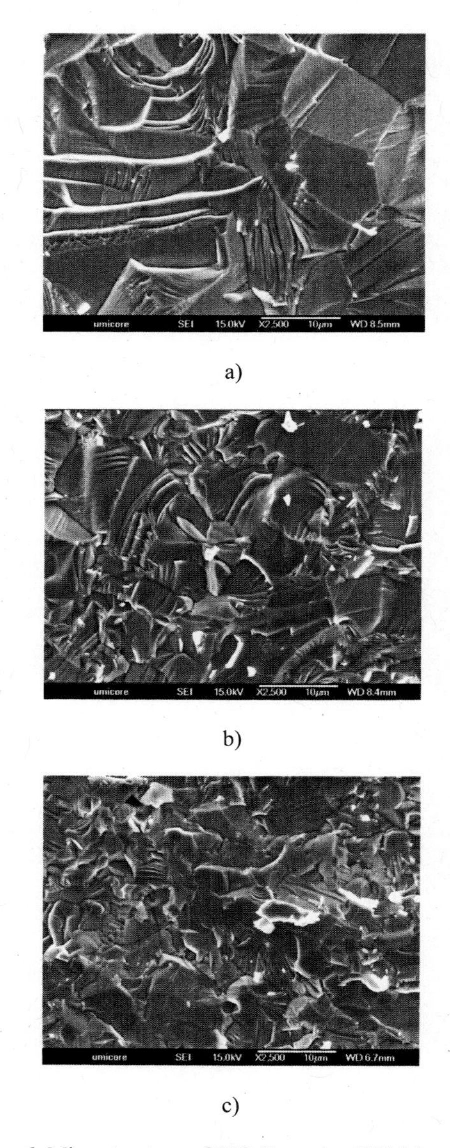

Fig. 6. Microstructure of ITO Ceramics (SEM Image)
a) 95/5; b) 90/10; c) 80/20

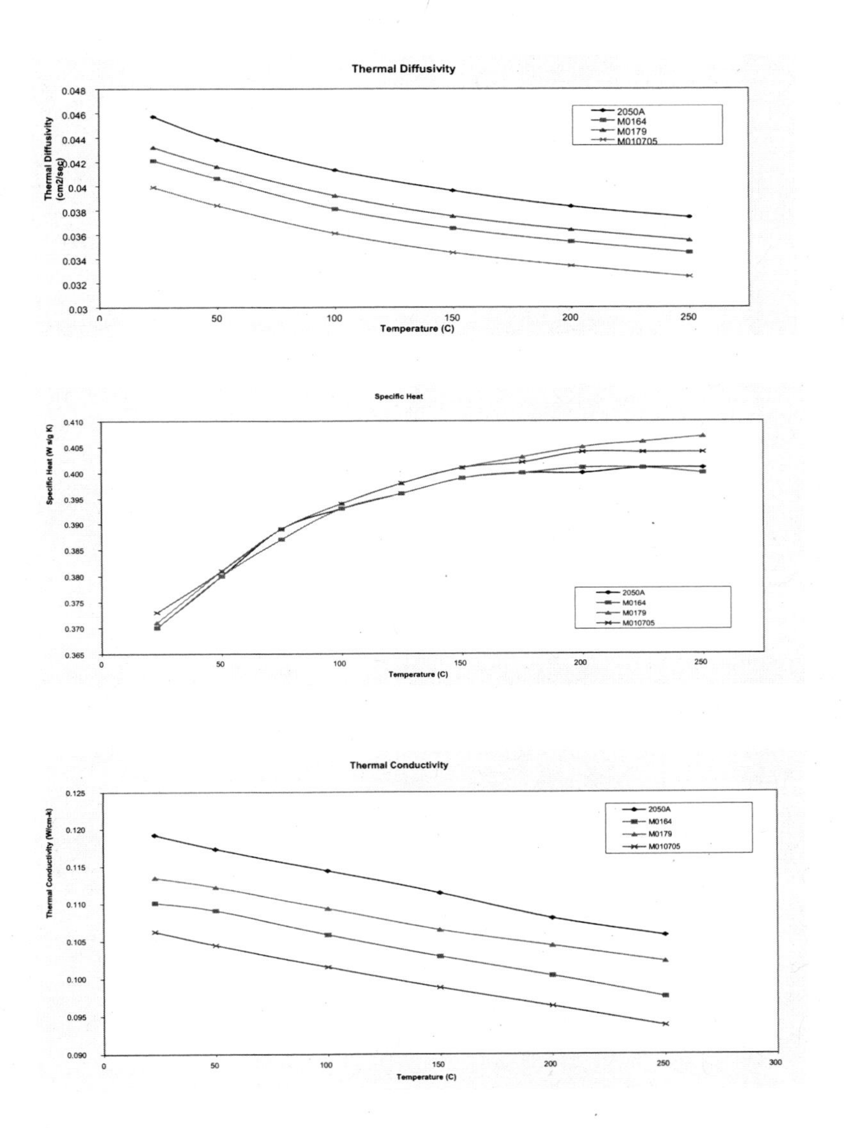

Fig. 7. Thermal Diffusivity, Heat Capacity and Thermal Conductivity of ITO 90/10 Ceramics

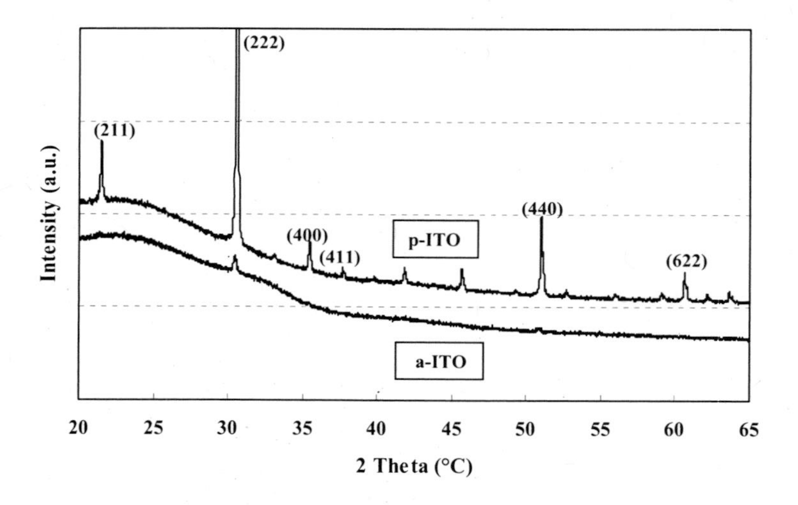

Fig. 8. XRD Spectra of ITO (90/10) Films Obtained at Different Sputtering Conditions (for Amorphous and Polycrystalline Films)

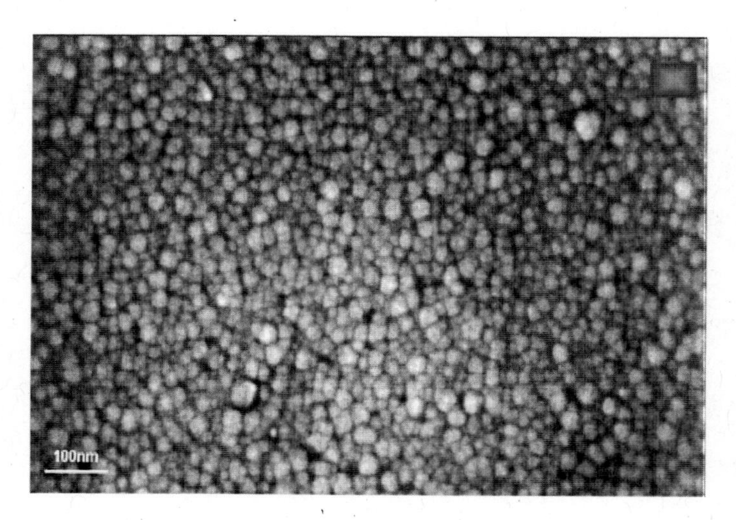

Fig. 9. Microstructure of ITO (90/10) Film Prepared by DC Magnetron Sputtering (Deposition at Room Temperature with Subsequent Annealing at 200°C during 1 hr.)

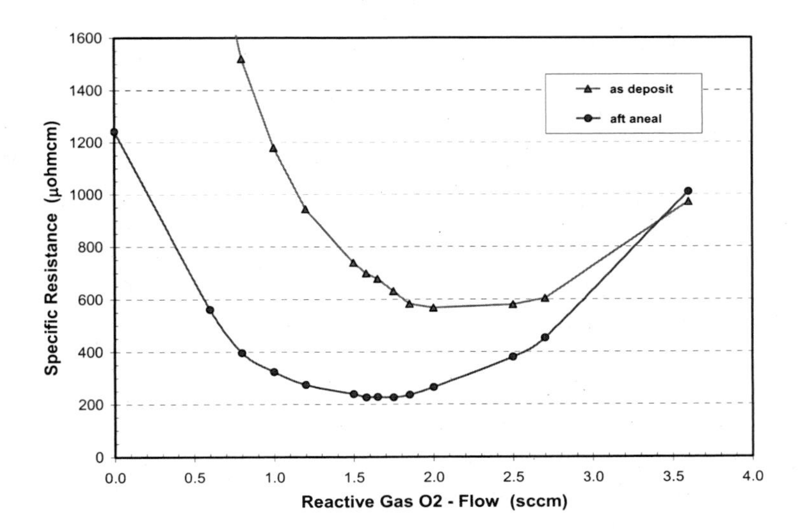

Fig. 10. Specific Electrical Resistivity of ITO Films Prepared from ITO 90/10 Ceramic Target by DC Magnetron Sputtering (after Deposition and after Annealing at 200°C during 1 hr.)

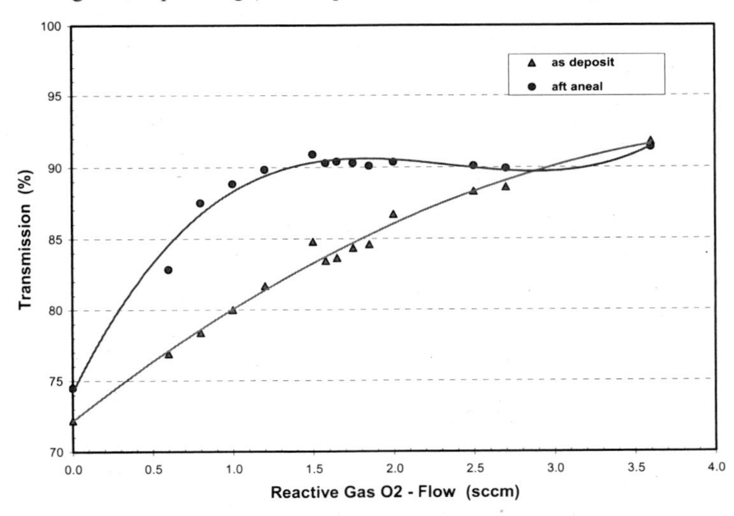

Fig. 11. Transmittance (at 550 nm Wavelength) of ITO Films Prepared from ITO 90/10 Ceramic Target by DC Magnetron Sputtering (after Deposition and after Annealing at 200°C during 1 hr.)

ENGINEERED OXIDE NANOFILMS PREPARED FROM SOLUTIONS AT RELATIVELY LOW TEMPERATURES

Arvid Pasto[1], Michael Pozvonkov[1], Morgan Spears[1], Evan Hyde[2], and Mark Deininger[1]
[1] C3 International, LLC, Alpharetta, Georgia 30004
[2] C2 Nano, Alpharetta, Georgia 30004

ABSTRACT

Engineered oxide nanofilms, prepared from solutions and heat-treated at up to about 450°C, are described. Films consist of layers of ca. 50-100 nm thickness, built up from oxide nanoparticles of 3 – 8 nm dimension. Most cations can be deposited, from a solution liquid which is tailored to this cation. The liquid attaches itself to a solid surface, and reacts with the native oxide on this surface. Upon heating, the solution molecules decompose and are eliminated, leaving the cation on the surface as an oxide. Heating also promotes an interaction of the cation, the oxygen atoms, and the elements in the substrate, resulting in an interdiffusion zone between the nanofilm and the substrate. Mixtures of solutions, or single solutions containing a mixture of cations, can be used, to form multi-cation oxides, such as yttria-stabilized zirconia, indium-titanium oxide, Gd-doped ceria, and so on.

These nanofilms offer numerous interesting properties, including coking resistance, carburization protection, corrosion- and erosion-resistance as "passive" behaviors, and are being developed for more "active" areas such as fuel cell electrodes and electrolytes, and solar photovoltaic receivers.

The process for depositing these nanofilms is simple, involving minimum substrate surface preparation; application of the liquid via any of a number of means (spraying, spinning, dipping, swabbing, etc.); and heat-treatment via simple furnacing, and/or induction, infrared, laser, or microwave heating.

THE C3 COATING TECHNOLOGY

C3 International, LLC ("C3") is the developer and owner of the patented Metal Infused Surface Treatment ("MIST") technology, which allows for in situ creation of metal oxides on metal and ceramic surfaces from nearly all metals of the periodic table. The MIST technology also allows for nearly any combination of any or all of those elements to create novel surface nanofilms that can resist fouling, erosion, corrosion, and many other types of damage mechanisms. Recent efforts by C3 staff have developed nanofilms for "active", rather than simply "passive" films, for use in solar photovoltaic cells, solid oxide fuel cells, and other electronic or ionic conducting applications.

Nanofilms developed by C3 have included stabilized zirconia (using various types of stabilizer cation), cerium oxide, aluminum oxide, combined alumina-silica, and others.

C3's technology has recently been recognized by the materials community in America as being extremely important. C3 was awarded an "R&D-100" award in 2006 as one of America's most important new technologies, and in 2007, the technology was further recognized by being awarded a "Nano-25" award as one of the 25 most significant nano-materials innovations of the year. C3 is the first nongovernmental agency to open office space in a U. S. Department of Energy ("DOE") sponsored lab, Oak Ridge National Laboratory ("ORNL"). As such, we have been working very closely with the scientists there to characterize our novel materials and understand their functional mechanisms.

Scanning Auger nanoprobe studies (Figure 1), along with transmission electron microscopic analyses (performed at ORNL) of C3's nanofilms (Figures 2 and 3), show an interdiffusion region where the coating layer material has diffused into the substrate, significantly increasing the bond strength and reducing the risk of its spalling off the surface.

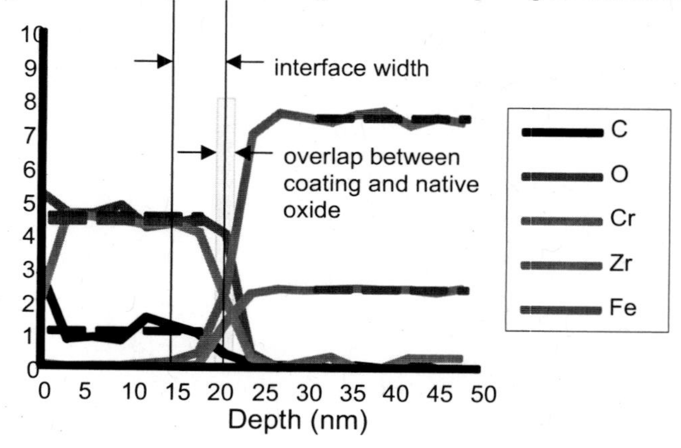

Figure 1. Scanning Auger nanoprobe results on a C3 zirconia film deposited on steel.

Figure 2. Transmission electron micrograph of a C3 zirconia film deposited on steel. Careful observation will show that there are three layers, resulting from three applications.

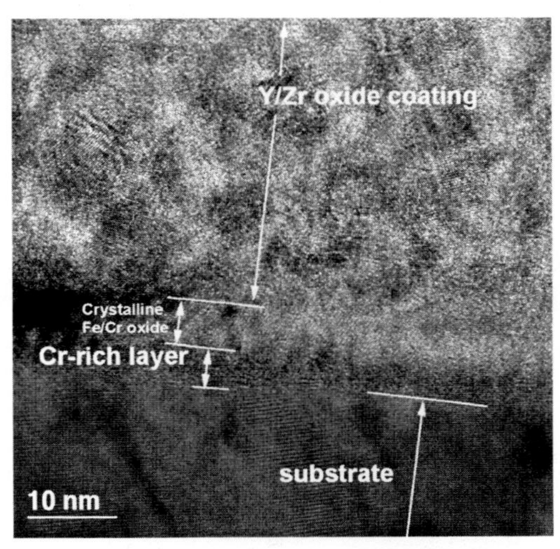

Figure 3. Transmission electron micrograph of a C3 zirconia film deposited on steel. Note the reaction and interdiffusion layer under the film.

Scientists at ORNL noted the crystallites created by our process were 3 to 8 nanometers in dimension. This finding was verified by synchrotron x-ray diffraction analysis performed through the High Temperature Materials Laboratory's (HTML) User Program, with diffraction being performed at the National Synchrotron Light Source at Brookhaven National Laboratory (Figure 4). We believe this small particle size results in extremely strong and dense films, while the interdiffusion region allows for inter-mixing of alloy and film constituents, and subsequently increases the bond strength between the alloy and the engineered nanofilm.

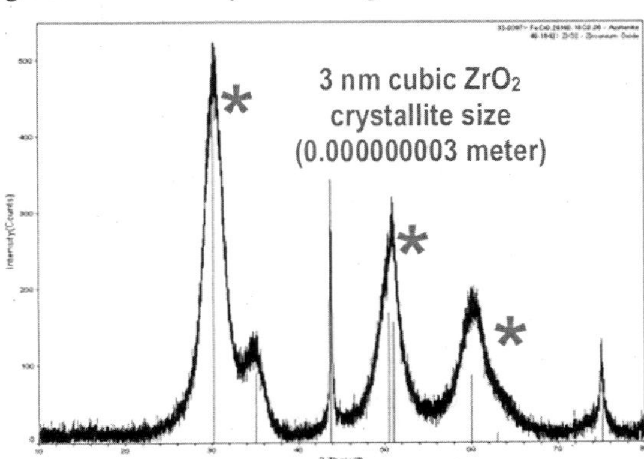

Figure 4. Synchrotron x-ray diffraction pattern of C3 zirconia nanofilm.

Nanofilms have been developed to resist corrosion in several industrial applications. One particular film has been developed which is extremely resistant to aqueous-based corrosion, including aqua regia. (See Figures 5 and 6 below) This same film has been proven to be resistant to elevated temperature salt-spray corrosion.

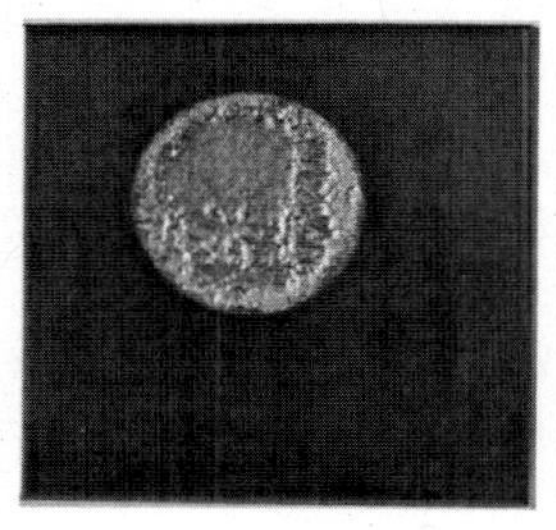

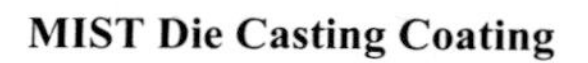

MIST Die Casting Coating **MIST Corrosion Coating**

Figure 5. Aqua Regia Test. Sample exposed to specific volume of 3 parts HCl to 1 part HNO3 for 1 hour. Dark region on left specimen is coating; light region is corrosion area, where film has been eaten through.

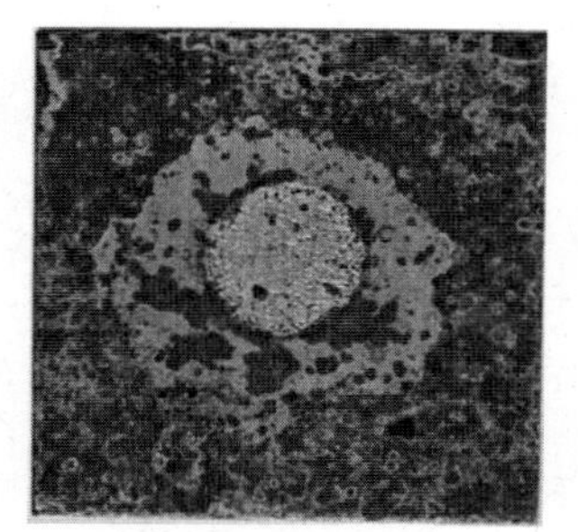

MIST Die Casting Coating **MIST Corrosion Coating**

Figure 6. High Temperature Salt Spray Test. Samples heated to 900°C in air, then periodically sprayed with 10% salt water solution for 1 hour. Circular region on left specimen is corrosion through the film.

These films are deposited in a very simple fashion: the metallic surface is prepared by simple cleaning and/or surface finishing techniques; a low-viscosity liquid precursor containing the cation(s) of interest is applied to the metal surface, by spraying, dipping, spinning, swabbing, or any other technique for applying a liquid to a surface; the metal part or component so coated is heated in air or inert gas atmosphere to about 450°C to remove the hydrocarbon backbone of the precursor and leave behind the cation as an oxide film.

The precursor compound is an organic molecule which is selected to exhibit reactivity to the metal surface, and which has the desired cation attached to the other end. After application of the compound to the metal surface, simple heating in air decomposes the organic molecule and results in a thin nanofilm of the cation as an oxide. Precursor molecules are described in great depth in the various C3 patents. This process typically leaves a dense nanofilm, with a few pores and/or cracks, depending on the rate of organic decomposition, thermal expansion mismatches, etc. The film is so thin that it offers no thermal resistance to speak of, and its thermal properties are unknown.

The technology is also unique in its ability to lay down nanofilms in-situ, onto very large systems, including those where line-of-sight access is not available. Many other coating technologies require vacuum, or at least controlled atmosphere, equipment, which greatly limits the size of component that can be coated. These expensive processes also, of course, increase cost of the coating. C3's lack of need for line-of-sight access, as required for processes like physical vapor deposition or sputtering, is another plus for C3's technology. The low viscosity liquid can be applied in various ways to any size or complexity of shape component. For example, in February of 2009 a carbon-fouling-resistant C3 coating was applied to the INSIDE of 970+ (continuous) meters of 76+ mm diameter tubing, in-situ, at a petroleum refinery in Wyoming. In June 2009, another refinery was a test site, which required the coating of the interior of over 3050 m of tubing (see following section).

One of the potential problems of preparing these nanofims on metallic surfaces arises from the temperature currently required to produce the oxide nanofilm (450°C). This temperature may be too high for some metals, or for alloys which have been heat-treated to develop specific structures or stress states, e.g., levels of martensite in steels, or tempering schedules applied. C3 is working to evaluate newer potential precursor molecules and heating techniques, which will either allow lower temperature final heat-treatment steps, or allow use of rapid surface heating of the substrate-chemical deposit combination in such a manner as to not alter the structure/stress state of the underlying alloy. It is already known that these engineered nanofilms can be heat-treat cured by use of rapid infra-red heating techniques, wherein the surface of the alloy is barely changed while the coating matures. Similarly, surface heating of the alloy by e.g., induction may allow the bulk temperature of the component to remain unchanged while the coating cures. Laser heat-treatment curing can also be utilized, as could microwave heating of the film precursor.

COKING AND CARBURIZATION RESISTANCE EXAMPLES

The major application for C3's zirconia nanofilm is in the petroleum refinery operation called "coking".

Petroleum Refining Industry Overview

Oil refineries in the United States are operating at increased production capacity in order to keep up with consumer demand for gasoline and diesel fuels, which continues to increase in spite of recent economic and environmental developments. Worldwide consumer demand is being fueled by a number of factors, including the dramatic economic expansion and development that is ongoing in China and India, which have increased both drivers and transportation of goods.

As demand grows, the supply of easy to refine "light sweet" crude oil is declining, forcing the oil industry to utilize larger percentages of "heavy" and "sour" crude oil. The cost to retrieve and process this lower-grade ("heavy sour") crude oil is significantly higher than that of the higher quality ("light sweet") crude oil that the industry has relied on for the past century. Increasing quantities of crude oil must be processed by a "delayed coker" in the refining process, upgrading it into fuel. While this "coking" process has existed for over 70 years, declining crude oil quality and increasing end product demand is resulting in increased usage for this delayed coking process.

Fundamentally, all process heaters suffer from some level of fouling accumulation, which results in decreased heat transfer rates. To achieve the same throughput and conversion at decreased heat transfer rates, the corresponding result is for the process heaters to burn more fuel, which represents in a major inefficiency. Fouling mitigation, which is the core of the Coker Coaters, LLC (a subsidiary of C3 International, LLC) technology, can have a significant effect in reducing the emissions in these largest emitters.

Fouling occurs at varying rates for different types of process heaters. Typically the larger the molecules being processed or the higher the severity (or temperature) of the process, the higher the fouling rate. For this reason, C2 has chosen its initial focus to be on the highest fouling processes in the sector to achieve the largest benefits. These are the delayed coker furnaces. As the technology matures, other process heaters including ethylene cracker furnaces and heat exchangers will be considered to further improve the benefits.

Delayed Coking Process

Delayed coking is a thermal conversion process which is heavily dependent on heat input from hydrocarbon fired furnaces. Coker heaters are directly affected by the heavier crudes that they are now having to process. Heavy crudes contain long hydrocarbon chains that must be heated to high temperature in order to be turned into more useful products. Heating the oil until it separates at a molecular level is called thermal cracking. This is done by running the heavy oil, called "resid" through approximately 1100 m of metal tubes in a very large heater, which heats the oil to about 480 – 510°C.

The problem of fouling within delayed coker heaters is that coke attaches to the inside walls of the heating tubes rather than traveling downstream to be removed in the coke drum. Coke is an effective insulator and its buildup inside the heater tube is measured in part by the tube temperature increase that is necessary to maintain the temperature required for the thermal cracking.

The petroleum refining industry and its equipment and materials suppliers have been grappling with the coking problem for decades. Papers have been published in numerous venues describing the phenomenon,[1,7] chemical additives to the resid, new techniques for pigging, and, of course, numerous coating techniques[2,3] and new tubing or piping materials.[4-6]

The coke must be cleaned out. One method is a process called mechanical pigging. Pigging involves shooting a semisolid cylinder with tungsten carbide teeth through the tubes, where these teeth tear off and clean out the coke. On average, this cleaning process typically takes four days per shutdown, and is required every three to 12 months, depending upon the type of oil used at the refinery and the process conditions. There are about 200 delayed coker heaters in North America and over 375 worldwide.

Through mitigation of the fouling via application of C2's patented MIST technology, coker heater fuel consumption at constant throughput can be reduced by as much as 2-3%.

The fouling issues with delayed cokers are well understood. Industry has been searching for mitigation techniques for these services for decades. C2 has passed a critical test for the process in an independent laboratory (see following) and is working towards field application and commercialization with a number of industry majors. This test is a modified Alcor test for residual oil fouling which is used to characterize novel materials and additives in an accelerated ninety minute fouling test at 540°C. This test has been used in the petrochemical industry for more than twenty years. The results have been compared to prior materials and chemistry and have shown remarkable results.

The test, performed at F.A.C.T. in Houston, by Dr. Ghaz Dickakian, involves circulating residual oil over tubes of standard petroleum refinery alloy, which contain internal heaters held at constant power level. The temperature of the oil is measured versus time, and, as coke build up on the alloy heater tubes, the temperature of the oil decreases. Several compositions of C3 nanofilms were tested in this apparatus (results below).

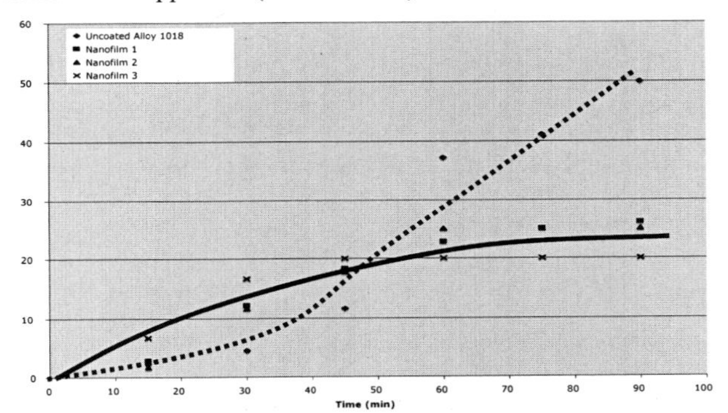

Figure 7. Temperature decrease versus time for residual oil flowing over heater tubes being maintained at a constant internal temperature (543°C). Note that the uncoated tube becomes coated with coke and heat transfer is reduced, while all of the C3 coating compositions reach a plateau at which further temperature decay is prevented (e.g., heat transfer is maintained because of lack of fouling by coke).

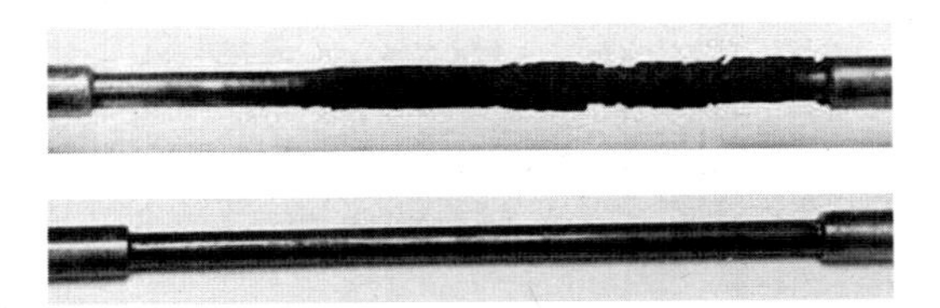

Figure 8. Uncoated (top) and C3 nanofilm coated (bottom) alloy tube exposed to residual oil on the outside, while being heated by heaters inside the tubes. Note that the C3 coated tube has much less, and more uniform, coke buildup.

A second (potential) major application operation called "ethylene cracking". Carbon deposit fouling and carburization cracking of steam or ethylene cracking furnace tubes are portions of the most severe fouling service in all petrochemical applications. Total worldwide production of ethylene and other basic petrochemical monomers via this process was 119.6 million tpy as of Jan. 1, 2008, with an expected growth of 2% per year for the next five years. Ethylene is a key basic chemical element in the production of many types of plastics, surfactants, detergents, and agricultural ripening agents. The critical step in the production of this product is the rapid heating of a feedstock, less than 1 second to more than 900° C, in a furnace. Carbon buildup, or fouling, can limit the production runs to as short as a few weeks before cleaning is required. The cracking issue is a large safety concern and requires the producers to routinely replace equipment and use exotic alloys where possible.

An experiment was performed in the Corrosion Science and Technology Group of the Materials Science and Technology Division of ORNL, to simulate ethylene cracking conditions. Alloy samples, with and without a C3 rare-earth element (REE)- stabilized zirconia nanofilm, were exposed to a slowly flowing 99% H_2/1% CH_4 atmosphere at 900°C for times up to 1000 hours. Complete test details and results are being prepared for a future paper, so only a summary of the coking and carburization results will be presented here.

Figure 9, below, shows the measured weight gains for the typical petroleum refinery tubing alloy, I803, with and without the C3 nanofilm. Here the weight pickup of the coated steel sample is seen to vary from about 80% less at 200 hours to about 50% less at 1000 hours.

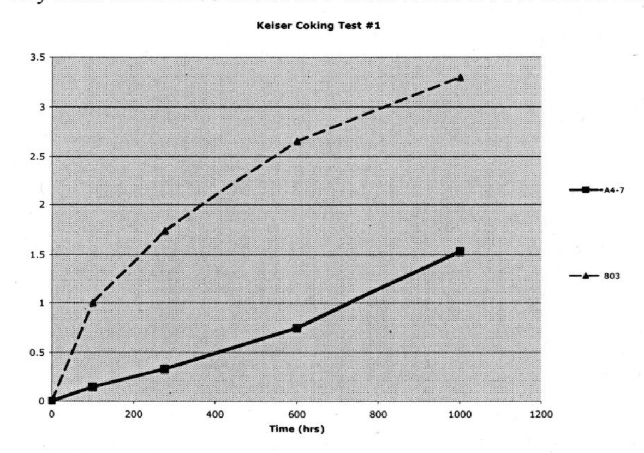

Figure 9. Coking and carburization test results. Weight gain for the uncoated I 803 steel coupon is significantly greater than for the coated coupon.

Most of this weight gain is from carburization, as will be shown later, but some coke deposition is also evident. SEM examination of the tested coupon surfaces showed copious carbon deposits.

Specimens were mounted in epoxy, cut into two portions across their diameters, and polished for electron microprobe analysis. Several analyses were performed as described earlier. A BSE image of the uncoated alloy I803 specimen is presented in Figure 10, along with corresponding wavelength-dispersive x-ray analytical elemental photographs.

The microstructure of the alloy has been radically altered by the exposure to the methane/hydrogen mixture, with large amounts of carbon having been absorbed and converted into chromium carbides both in the grains and grain boundaries.

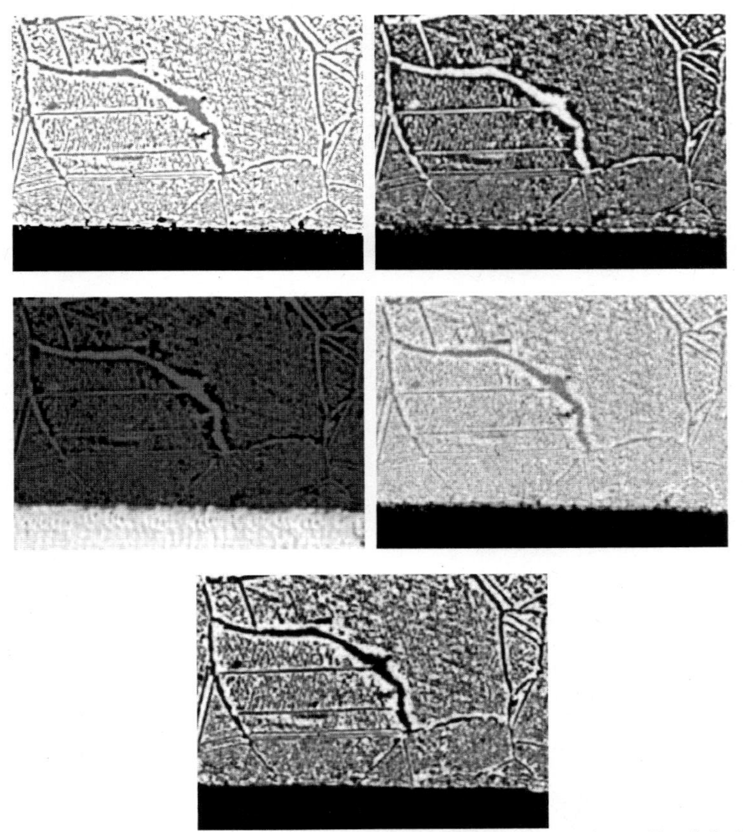

Figure 10. Electron microprobe results for uncoated alloy I803 specimen. Top left: BSE image. Top right: Cr image. Middle left: C image. Middle right: Fe image. Bottom: Ni image.

Electron microprobe analyses of the stabilized zirconia-coated specimen yield similar results: chromium carbides are found throughout the alloy internal structure (Figure 11), and chromium is found at the specimen surface along with Zr, REE, and O.

Figure 11. Electron microprobe results for stabilized zirconia-coated alloy I803 specimen. Top left: BSE image. Top right: Cr image. Middle left: C image. Middle right: O-Zr-Cr image. Bottom: Zr-REE-Cr image.

In an attempt to determine the extent of carbon penetration into the specimens, four scans completely across each specimen, at locations roughly one-quarter of the distance from each end, were set up. The probe stopped at 52 locations and measured Cr, Fe, Ni, Si, and C. The entire data set is not presented here, but the following two figures and the table provide summaries.

First is illustrated one of the scans (the third of each set of four) from each of the uncoated alloy (Figure 12) and a coated specimen (Figure 13).

The uncoated alloy specimen shows high carbon levels throughout the specimen, while the coated specimen shows a much lower carbon level and more uniform levels of Cr, Ni, and Fe through the center of the specimen than the uncoated specimen.

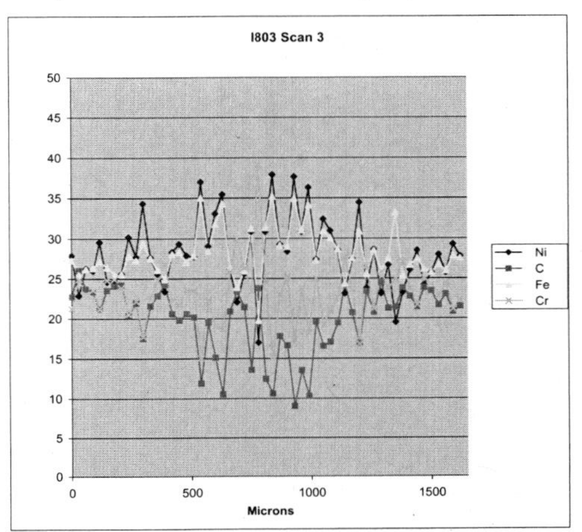

Figure 12. Microprobe results on an uncoated I803 alloy specimen exposed to 99% H_2/1 % CH_4 at 900°C for 1000 hours. Significant carbon pickup is noted, with nearly one atom in five in the structure being replaced by carbon.

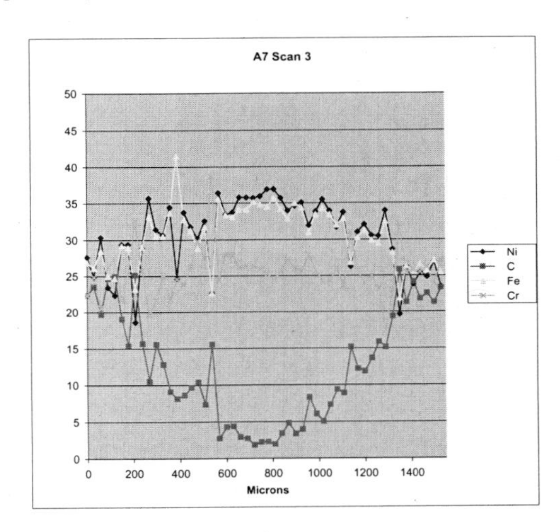

Figure 13. Microprobe results on a stabilized zirconia-coated I803 alloy specimen exposed to 99% H_2/1 % CH_4 at 900°C for 1000 hours. Carbon pickup is much reduced from the previous specimen.

Table 1 summarizes the probe data from these specimens, presenting the normalized atomic percent of the five elements averaged over the 52 counting locations along each of the four scans across the specimen.

Table 1. Electron Microprobe Data for Uncoated and C3-Coated Coupons

SPECIMEN #	X-Section #	Normalized Atomic Percent (Avg. of 52 Point Counts) Ni	C	Si	Fe	Cr
I803 Uncoated	1	26.9	21.1	0.8	27.2	24.1
	2	28.4	19.8	0.8	28.1	23.0
	3	28.0	19.9	0.8	28.1	23.1
	4	27.1	20.7	0.8	27.5	23.8
	Average	27.6	20.4	0.8	27.7	23.5
I803 Coated	1	32.5	8.8	1.0	31.7	26.1
	2	31.6	10.4	0.9	31.1	26.0
	3	30.5	12.4	0.9	30.5	25.6
	4	32.1	9.5	0.9	31.5	26.1
	Average	31.7	10.3	0.9	31.2	26.0

Table 2 illustrates the average carbon content (column 2) of the two specimens, then presents the percentage difference between uncoated I803 and the coated specimen (column 3). The nanofilm has reduced the carbon uptake by 50%.

The fourth and fifth columns present the average weight changes of the two materials, and the difference between the coated material versus the uncoated, respectively. The similarity in reduction in carbon uptake inside the specimens and the differences in total measured weight change indicate that the films were actually effective at preventing carburization of the alloy. The amount of coke buildup on the surface is likely a small contributor to the weight changes observed.

Table 2. Carbon Pickup of Uncoated and C3-Coated Coupons

Specimen	At % Carbon	Difference (%)	% Wt. Change After 1000 hr	Difference (%)
I803	20.4	50	3.3	53
A7	10.3		1.54	

CONCLUSIONS

Results of weight gain measurements, electron microscopy, and electron microprobe analyses of materials exposed at 900°C for 1000 hours in a flowing atmosphere of 99% H_2 and 1% CH_4 have shown that an engineered nanofilm of stabilized zirconia is effective at reducing carburization of an underlying alloy substrate. The film remains relatively intact, although it appears that chromium from the alloy has migrated outward through it, and carbon has migrated inward through it. It is not known at this point how the reduction in carburization of the alloy has affected its properties, such as strength and ductility. Further experimentation, perhaps using small tensile specimens, is warranted.

REFERENCES

[1]Asrar, Ashiru, and Al-Beed; *Erosion Damages of High Temperature and Wear Resistant Materials During Ethylene Cracking in a Petrochemical Plant*, NACE Paper 00087, Corrosion 2000

[2]Smith, Kempster, Lambourne, and Smith; *The Use of Aluminide Diffusion Coatings to Improve Carburization Resistance*, NACE Paper 01391, Corrosion 2001

[3]Redmond, Chen, Bailey, and Page; *A Low Coking and Carburization Resistant Coating for Ethylene Pyrolysis Furnaces*, NACE Paper 01392, Corrosion 2001

[4]Nishiyama, Semba, Ogawa, Yamadera, Sawaragi and Kinomura; *A New Carburization Resistant Alloy for Ethylene Pyrolysis Furnace Tubes*, NACE Paper 02386, Corrosion 2002

[5]Hendrix; *Comparative Performance of Six Cast Tube Alloys in an Ethylene Pyrolysis Test Heater*, NACE Paper 430, Corrosion 98

[6]Milner; *Carburization of Emerging Materials*, NACE Paper 269, Corrosion 99

[7]Nishiyama and Otsuka; *Degredation of Surface Oxide Scale on FE-NI-CR-SI Alloys Upon Cyclic Coking and Decoking Procedures in a Simulated Ethylene Pyrolysis Gas Environment*; NACE Corrosion, Jan. 2005

EXPERIMENTAL STUDY OF STRUCTURAL ZONE MODEL FOR COMPOSITE THIN FILMS IN MAGNETIC RECORDING MEDIA APPLICATION

Hua Yuan and David E. Laughlin

Department of Materials Science and Engineering, Carnegie Mellon University, Pittsburgh, PA 15213, USA and the Data Storage Systems Center, Carnegie Mellon University, Pittsburgh, PA 15213, USA

ABSTRACT

Composite thin films are a new area of study with many applications, e.g. metal + oxide thin films in high density magnetic recording media. Engineering the processing, microstructure and properties of these thin films is of great importance. It has been found that microstructures of the thin films depend strongly on the oxide volume fraction and pressure during sputtering. Surface diffusion and self-shadowing effects are found to play important roles in determining various thin film microstructures under different processing conditions. Four characteristic microstructural zones could be generally distinguished: "percolated type"(A), "maze-like type"(T), "granular columnar type"(B) and "embedded type"(C). This modified structural zone model of composite thin films has been proposed as a supplement to Thornton's model for sputtered thin film system.

This research has been sponsored by Seagate Technology and the DSSC of CMU.

INTRODUCTION

Composite thin films e.g. metal + oxide thin films are widely used in high density magnetic recording media.[1-4] These films are composed of the crystalline metallic phase and a secondary amorphous phase. Engineering the processing, microstructure and properties of these thin films is of great importance. The studies about composite thin films are limited and non-systematic in literature. In addition, the real microstructures of composite thin films can be very complicated by varying various factors. Even a slight change of microstructure could cause a significant degradation of film properties. As a result, it is important to have a comprehensive understanding of how the composite thin films grow and how their microstructures evolve during sputter deposition, especially for composite thin films fabricated at room temperature for current perpendicular magnetic recording media. In order to predict the microstructures of composite thin films, it will be informative to construct a phenomenological structural zone model which includes the relative volume fractions of the two phases present. This supplements Thornton's structural zone models for the sputter deposited thin films. In this paper, the major results based on the composite Ru + oxide thin films for magnetic recording media application will be discussed in details.

It has been found that microstructures of the thin films depend strongly on the oxide volume fraction and pressure during sputtering. Surface diffusion and self-shadowing effects are found to play an important role in determining various thin film microstructures under different processing conditions. Four characteristic microstructural zones could be generally distinguished: "percolated type"(A), "maze-like type"(T), "granular columnar type"(B) and "embedded type"- a microstructure with metal nanocrystals embedded in the amorphous matrix (C). This modified structural zone model of composite thin films has been proposed as a supplement to Thornton's model for sputtered thin film system.

EXPERIMENT

Thin films: glass substrate \ Ta (3 nm) \ Ru (low pressure, 15 nm) \ Ru + TiO_2 (10 nm) \ CoPt + oxide (11 nm) \ carbon overcoat was sputter deposited at room temperate. Ta \ Ru seedlayers are utilized to provide good adhesion and (00.2) crystallographic texture for the composite Ru + TiO_2

layer. Ru + TiO_2 interlayer was studied at three pressures: 3 mT, 30 mT and 60 mT. The volume fraction of TiO_2 varied from 0 vol % to 50 vol % in the Ru + oxide interlayer. The magnetic layer in each sample was fixed with the same composition and thickness. Hysteresis loops were measured by the magneto-optic Kerr effect magnetometer (MOKE). The crystalline orientation was evaluated from X-ray rocking curve scans of Ru (00.2) peaks taken on a Philips X'pert diffractometer. Microstructures and surface morphology of samples deposited up to composite Ru + TiO_2 layer and samples up to composite CoPt + oxide magnetic layer were analyzed on the JOEL 2000 transmission electron microscope (TEM).

RESULTS AND DISCUSSION

By means of varying the oxide volume fraction and pressure during sputtering, the microstructure of composite Ru + TiO_2 thin films underwent a significant change. TEM bright field imaging technique has been used to observe the microstructure features of the thin films in the study. The morphologies of such thin films are schematically drawn in Fig. 1, where the dark area represents for the crystalline metal phase and white area represents for the amorphous secondary phase region (e.g. oxide). Four characteristic microstructural zones could be generally distinguished: "percolated"(A), "maze-type"(T), "granular"(B) and "embedded". This is also listed in Table I. Zone A represents the low pressure, low volume fraction (VF) region of the plot, where only the surface diffusion effect is crucial. A continuous metallic film with oxide phase of small feature sizes is commonly observed. The thin films exhibit good crystallinity. The qualitative texture evolution of the composite thin films as a function of oxide volume fraction and pressure is mapped in Fig. 2.

Zone B is for high pressure, low oxide volume fraction region, where the self-shadowing effect overwhelms the surface diffusion effect. A columnar granular microstructure is the typical type of morphology in this region. Within the medium pressure and low oxide volume fraction region, denoted as transition zone T, the thin film microstructure shows networking or maze-like characteristics. Both the surface diffusion and the self-shadowing effects play equivalent roles in determining the final thin film morphology. With an increase in the amorphous phase fraction, the feature size decreases continuously.

Zone C covers rest of the plot, which is within high oxide volume fraction region and under all pressure range. Here, the metal nanocrystals are dispersed and embedded within the amorphous oxide matrix.

Fig. 3 shows the plan-view TEM bright field micrographs of composite Ru + oxide thin films with 4 different microstructures. They are located at both low and high oxide volume fraction ends of zone A (percolated) and zone B (granular). It is obvious that the structure zone plot describes well the microstructures of the composite thin films (Ru + TiO_2) sputter deposited at different pressure and oxide volume fraction values. Region 1 shows a columnar granular microstructure of Ru grains with voids and oxide rich grain boundaries under high pressure and low oxide volume fraction condition. This is dominated by the enhanced self-shadowing effect. Region 2 shows a dense Ru grain structure with low defects density, low voids density and smooth surface. This is mainly determined by the high surface mobility induced by the low pressure and low oxide volume fraction condition. As for region 3, under low pressure and high oxide volume fraction condition, the oxide phase forms in the metal thin film and it shows a percolated microstructure. Fine Ru and oxide grains as small as 2 ~ 3 nm coexist with up to 40 vol % oxide addition. The density of the oxide phase increases with increasing oxide amount. Region 4 displays a microstructure of many smaller columnar Ru grains with oxide segregated within the grain boundaries, compared to the microstructure of region 1. While the grain size becomes smaller, the grain uniformity degrades with increasing oxide amount.

Fig. 4 shows the plan-view TEM bright field micrographs of Co-alloy + oxide thin films that have been deposited on top of composite Ru + TiO_2 thin films with the 4 different microstructures shown in Fig. 3. With increasing oxide volume fraction in the composite Ru + TiO_2 thin films, the

CoPt grain size decreases. This demonstrates that the microstructure evolution in the composite underlayer affects the microstructure of the magnetic layer. As for region 1, the Ru grain size and CoPt matches well, implying a one-to-one epitaxial growth of the grains in these two adjacent layers. It is not surprising that the magnetic grain size does not change much on top of region 1 and region 2 microstructures, as not much oxide was added in the Ru layer. However, the CoPt grain size on top of Ru + TiO_2 thin film (region 3) of percolated microstructure has been reduced to 5.8 nm. The CoPt grain size on top of Ru + TiO_2 thin film (region 4) of smaller granular non-uniform microstructure has been also reduced to 6.6 nm.

In addition to the effect of the microstructure of the composite Ru + TiO_2 thin films on the microstructure of the magnetic layer, there is also a great effect on the resulting properties of the magnetic layer. Fig. 5 shows the perpendicular MH loops of Co-alloy + oxide thin films on top of composite Ru + oxide thin films with 4 different microstructures. Due to the perfect one-to-one epitaxial growth for magnetic film on top of region 1 underlayer, the films show the highest coercivity (H_c), the most negative nucleation field (H_n), a low (H_c-H_n) value and a high squareness value. This indicates a good inter-granular exchange-decoupling, crystallographic orientation and narrow c-axis distribution in the magnetic layer.

As for the magnetic film on top of the region 2 underlayer, although the microstructure of the magnetic layer is similar to the former case, the grains are highly exchange coupled due to the dense Ru grain structure. This decreases H_c and (H_c-H_n) values. However, squareness of the magnetic film is improved greatly due to the effect of very good crystallographic texture of the composite Ru underlayer, which was deposited under very low pressure and with very low oxide addition.

As the magnetic film is on top of region 3 underlayer, the very small magnetic grain size has decreased the H_c value. And the degraded texture of the magnetic layer resulting from the percolated underlayer microstructure reduces the squareness.

The magnetic film which grew on top of the region 4 underlayer, the magnetic grain size slightly decreases to 6.6 nm. It is likely that the granular microstructure in the Ru + TiO_2 underlayer is too non-uniform to form a one-to-one epitaxial growth. Therefore, both the smaller magnetic grain size and non-ideal growth decreases the H_c. Nevertheless, due to the segregation of high percentage of oxides in the grain boundaries, the magnetic grains are better exchange decoupled, judging from the hysteresis loops.

CONCLUSION

Metal + oxide (e.g. Ru + TiO_2) composite thin films were chosen as underlayer materials for magnetic recording media due to their unique segregated microstructure. The metal and oxide phases in the composite thin films are immiscible. The final microstructure of the interlayer depends on factors, such as, sputtering pressure, oxide species, oxide volume fraction, thickness, alloy composition, temperature etc. Moreover, it has been found that the microstructure of the composite thin films is affected mostly by two important factors - oxide volume fraction and sputtering pressure. The latter affects grain size and grain segregation through surface-diffusion modification and the self-shadowing effect. The composite Ru + oxide interlayers were found to have various microstructures under various sputtering conditions. Four characteristic microstructure zones can be identified as a function of oxide volume fraction and sputtering pressure - "percolated"(A), "maze"(T), "granular"(B) and "embedded" (C), based on which, a new structural zone model (SZM) is established for composite thin films. The evolution of microstructures determines the properties of the thin films significantly.

ACKNOWLEDGEMENTS

The authors would like to thank the financial support by Seagate Technology and Data Storage Systems Center (DSSC) at Carnegie Mellon University. We are grateful for discussions with Dr. Bin Lu, Dr. Yingguo Peng and Prof. Jimmy Zhu.

REFERENCES
[1]T. Oikawa, M. Nakamura, H. Uwazumi, T. Shimatsu, H. Muraoka, and Y. Nakamura, *IEEE Tran. Magn.*, **38**, 1976 (2002).
[2]H. Uwazumi, K. Enomoto, Y. Sakai, S. Takenoiri, T. Oikawa, and S. Watanabe, *IEEE Trans. Magn.*, **39**, 1914 (2003).
[3]Y. Inaba, T. Shimatsu, T. Oikawa, H. Sato, H. Aoi, H. Muraoka, and Y. Nakamura, *IEEE Trans. Magn.*, **40**, 2486 (2004).
[4]J. Ariake, T. Chiba, and N. Honda, *IEEE Trans. Magn.*, **41**, 3142 (2005).

Table I: List of four characteristics structural zones of sputtered composite metal + oxide thin films.

Zone	Pressure (mT)	Oxide volume fraction (0 – 100 %)	Structural characteristics
A	low	low - medium	"percolated"
T	medium	low - medium	"maze"
B	high	low - medium	"granular"
C	all	high	"embedded"

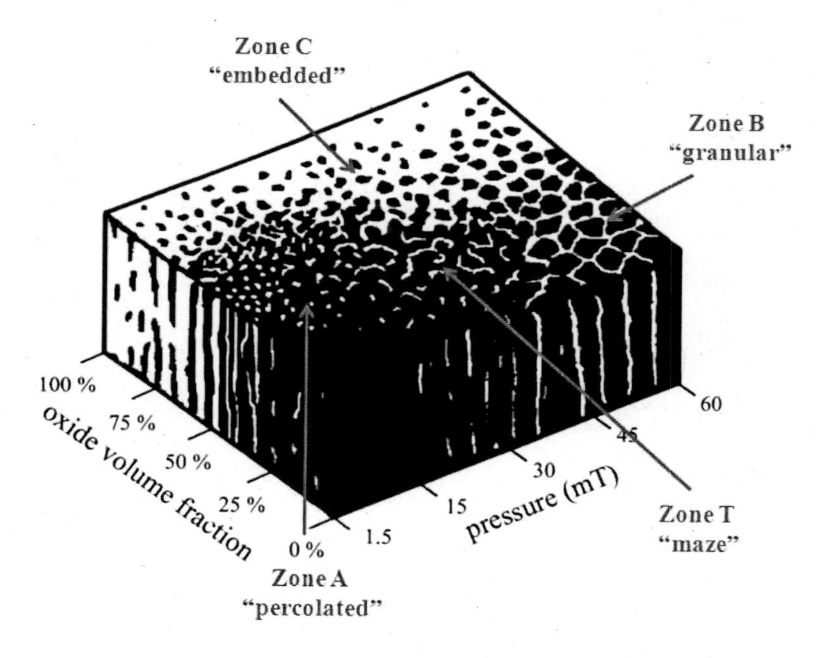

Figure 1 Schematic illustrations of the structural zones of composite metal + oxide thin films as a function of inert Ar gas pressure and secondary phase volume fraction (e.g. oxide volume fraction). Four characteristic zones – "percolated", "granular", " maze" and "embedded" are distinguished.

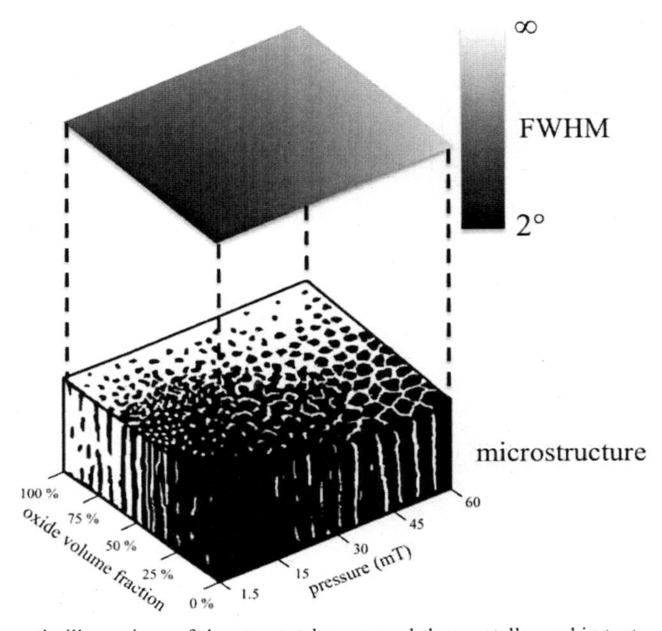

Figure 2 Schematic illustrations of the structural zones and the crystallographic texture of composite metal + oxide thin films as a function of inert Ar gas pressure and secondary phase volume fraction (e.g. oxide volume fraction).

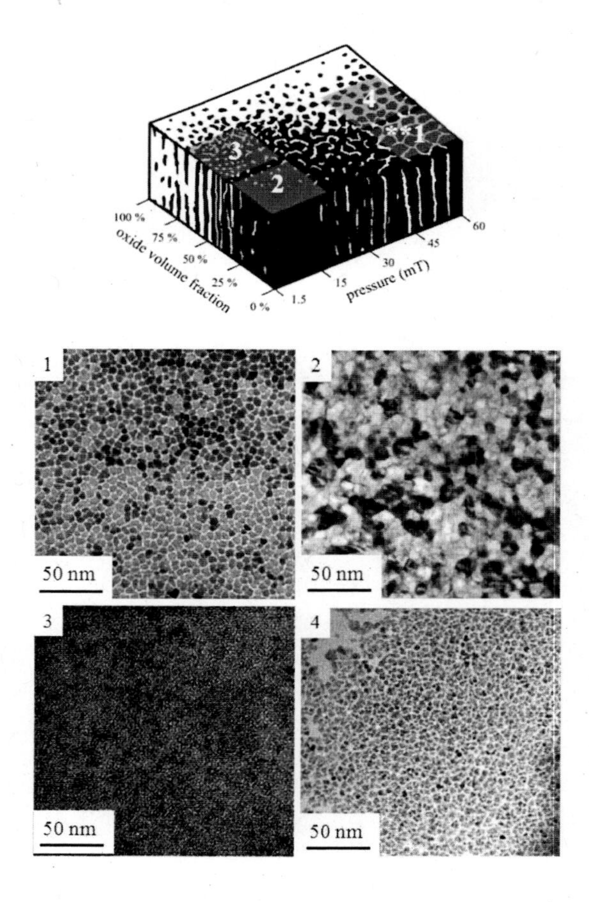

Figure 3 Plan-view TEM bright field micrographs of composite Ru + oxide thin films with 4 different microstructures.

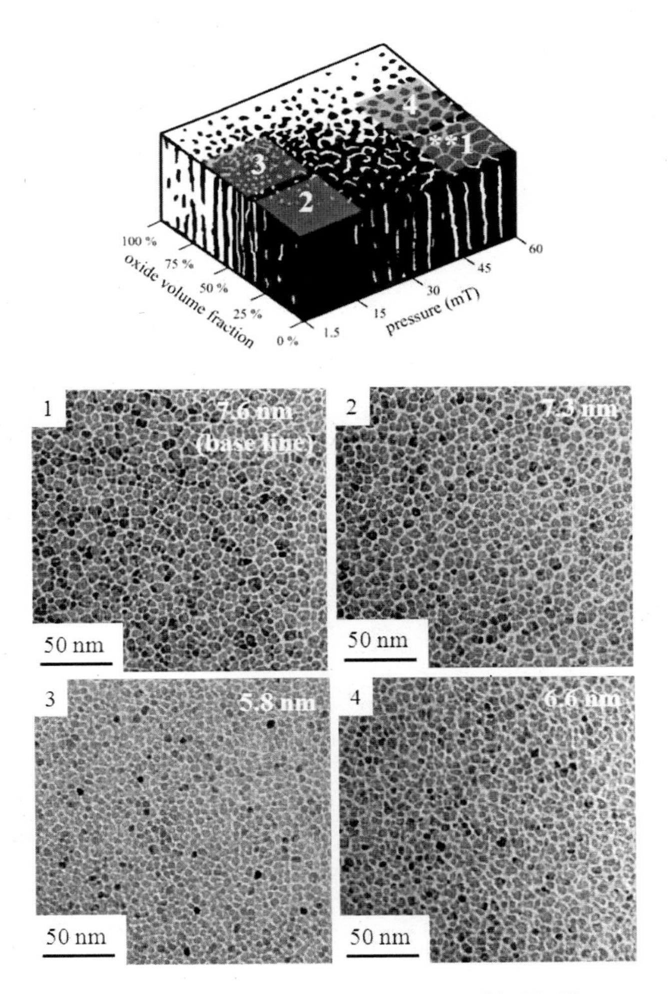

Figure 4 Plan-view TEM bright field micrographs of Co-alloy + oxide thin films on top of composite Ru + oxide thin films with 4 different microstructures.

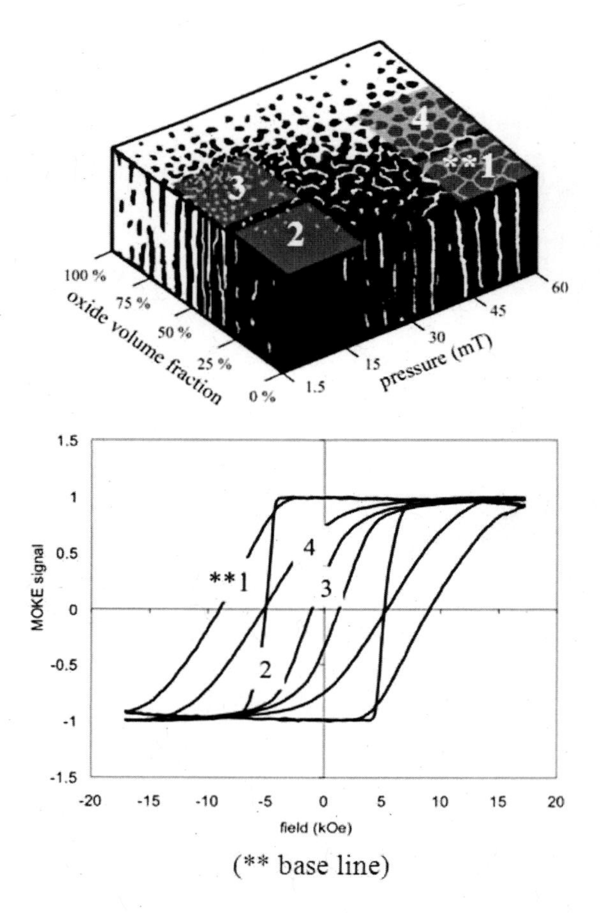

Figure 5 Perpendicular polar Kerr loops of Co-alloy + oxide thin films on top of composite Ru + oxide thin films with 4 different microstructures.

PROCESS OPTIMIZATION OF ION PLATING NICKEL-COPPER-SILVER THIN FILM DEPOSITION

Mike Danyluk
Mechanical Engineering PhD Student
University of Wisconsin Milwaukee
Milwaukee, Wisconsin

ABSTRACT

Simulation of thin film nano deposition manufacturing processes can give insight into process behavior and promote efficient resource allocation. If the process involves feedback control, (i) one can study controllability and observability of the system, (ii) know ahead of time what type of control scheme will be needed and (iii) the size and type of mechanical hardware to specify. DC ion plating systems are a collection of individual control systems that are run in either open loop or feed forward as a process. Arc disturbances from contamination or high voltage isolation breakdown can change process pressure by as much as 5mTorr and concurrently result first in a plasma voltage decrease followed by a voltage increase for up to 5 sec using PID control. Properties of the film can be significantly altered compared with previous layers resulting in significantly different failure modes of the film when tested in rolling contact fatigue. Ansi T5 ball bearings are coated with approximately 10nm of nickel-copper followed by 110nm of silver using an ion plating process. An ultra high vacuum rolling contact fatigue test platform is introduced for solid lubrication testing. High voltage events shorten test life and introduce a unique failure mode when tested in rolling contact fatigue. Voltage and pressure changes result in significant thickness monitor measurement errors. A regulator control scheme that minimizes process pressure overshoot is simulated.

INTRODUCTION

Arc discharge events resulting from contaminants during a deposition process are unavoidable in a manufacturing setting. Simulation of DC ion plating requires knowledge of mechanics, electronics, chemistry, and quantum mechanics. Simulation of ion plating and ion implantation at the sub element level has been explored extensively in the past two decades and has enabled accurate process modeling and simulation.

DC ion plating machines are a collection of closed loop subsystems that are usually run in open loop as a process. The voltage source, deposition sources, pressure monitor, and substrate manipulator operate independent of each other at the subsystem level. It is common practice that each system is started at a set point in time as prescribed by the process recipe, or reference input, and proceeds until stopped without process feedback from one another.

ION PLATING

Ion plating is a momentum transfer process in which the kinetic energy of the active gas ions is transferred to the deposition material through atomic ballistic collision. In this process argon ions collide with neutral atoms, silver for example, that have entered the plasma from either an evaporation or sputter source. The energy transferred from the argon ions to the silver atoms is sufficient to implant these atoms into the lattice structure of the substrate material and is related as,

$$\frac{E_t}{E_i} = \frac{4m_i m_t}{(m_i + m_t)^2} \cos^2 \theta \tag{1}$$

where E_i and E_t are the incident and target kinetic energies of the colliding ions and atoms, θ is the angle of the collision with m_i and m_t as the incident and target mass, respectively.

There are competing sputter removal and deposition mechanisms inherent with the ion plating process. The same argon ions that transfer momentum to the floating silver atoms will also displace already-deposited silver atoms from the substrate surface. [1]Mattox reports that when a large number of substrates are to be plated, ball bearings for example, the ion plating process has the added benefit of ion bombardment surface cleaning followed immediately by implantation and coating.

CURRENT RESEARCH

Simulation of doping profile formation and ion implantation are used in the manufacture of silicon based electronic devices. [2]Zechner et al. show that simulation based design has reduced development costs by 20 percent in 1999 and by almost 40 percent in 2008. Simulation of dopant profiles provides the designer with a tool for predicting spatial distribution of the dopants in the substrate that is energy dependent.

Plasma-target interaction modeling enables predictability of dopant distribution and subsequent film growth. [3]Moller et al. simulated a reactive magnetron sputter deposition process of TiN using a combination of simple models describing the plasma and first order surface/plasma interactions. [18]Malyshev et al. report that the density and structure of deposited Ti-Cr-V getter film was significantly influenced by process pressure and surface conditions. In a DC sputter process the degree of interaction between the plasma and target is controllable through process gas pressure and voltage.

The DC plasma can be interrupted by either contamination or high voltage isolation collapse during the process. Both of these interruptions effect deposition microstructure and film stress. The more severe of these interruptions is from voltage isolation breakdown and is critically linked to preventative maintenance schedules, cost, and chamber usage history.

[4]Qiu et al. use a particle method self consistent plasma model and cite limitations of fluid and kinetic plasma models. They use a 300VDC process at 0.4Pa and calculate an electron velocity of 10^7 m/s and collision frequency about 10^8/sec. Sheath thickness decreases with increase of both cathode voltage and gas pressure. Their model accounts for ionization, elastic collision, and excitation. They assume the energy of secondary electrons from the substrate is about 4eV and energy due to electric field is about 100eV.

[5]Moyne et al. presents efficacies of run-to-run process control, the so called least equipment invasive control scheme. Run-to-run control is a discrete process for machine control in which the product recipe is adjusted from run to run to meet quality/design intent needs. It is one way to correct for and prevent effects of process drift and shift based on process reference models and optimization targets. Control problems usually result from poor process models and poorly tuned controllers.

[6]Chan et al. use a continuum model that includes implantation, sputtering, and defect diffusion input to investigate ion irradiation effects on a thin Cu film. They use an optical measurement technique to detect curvature in the substrate during ion bombardment caused by film stress. They observe an immediate compressive stress during ion irradiation that reaches steady state after 500sec then switches immediate to tensile stress when sputtering is turned off.

Their experiments show that point defects generated by the ions cause film stress that is about 10-20nm deep for the implantation energy in their tests. Historically some researchers used molecular dynamics to conclude ion implantation alone caused stress, but Chan et al. report that molecular dynamics neglects recombination of vacancies and interstitial defects.

[19]Morley et al. use a cosputter-evaporation technique to control the amount of Fe and Ga in sputtered a film. A dc magnetron sputtering system is used to deposit Fe and a resistance heater crucible is used to evaporate the Ga. The plasma power and process pressure are used to effect change of the availability of Fe and Ga and thereby allow control of specific film constituent concentration.

[7]Kadera et al. use a cathodoluminescence method to measure residual stress within an interlayer dielectric film. They report successful detection within 50nm spatial resolution. [8]Bao et al. use H2, N2, N2/H2, O2, and Ar plasma treatments on a low k film and conclude that in low k materials plasma damage involves physical and chemical processes with the physical effects being dominated by the largest mass ions at high plasma powers, and chemical effects being dominate with lower mass ions at lower plasma powers. The tests were conducted with nominal plasma parameters of 30sccm at 30mTorr and150W plasma power for about 20mins.

[9]Liston uses a rolling contact fatigue test (RCF) device to study spalling and coating adhesion and life properties of TiN and NBN coatings applied to M50 steel. A superlattice of alternating layers of TiN and NBN coatings is applied using an unbalanced Magnetron DC sputtering system. The calculated hertzian contact stress was 3.4GPa and 5.2GPa with a rotation speed of 3600rpm using an oil drip lubrication system without recirculation. A 2 to 10 times improvement of L10 life was observed for coating thicknesses of .25 to .5um and specific periodicities of TiN and NBN interlayers. Scanning electron microscopy (SEM) and secondary electron imaging (SEI) was used to assess the coating after RCF testing. One interesting outcome was that most of the coating flaked off within the first 20hours of the tests and it was the remaining islands of coating that improved L10 life. The damaged coating was subsequently removed from the ball bearing contact zone by the oil drip.

[10]Liu et al. use substrate attributes such as micro-hardness, surface finish, microstructure, and residual stress to predict RCF life of bearing steels. They report good agreement between test results and their model in areas of high compressive residual stress. [11]Felmetsger et al. use pressure control to effect change in residual thin film stress. Precise control of argon pressure to manipulate compressive stress formation is applied. Gas pressure is manipulated to insure film uniformity as well. Essentially, they regulate the flux of ions from the discharge to the substrate by controlling argon gas pressure.

OBJECTIVES OF THIS PAPER

The objective of this paper is to demonstrate the benefits of a single closed loop control scheme applied to a DC ion plating process and to compare results to a typical system related to arc disturbances due to high voltage isolation breakdown and impurities. Determination of critical process parameters is studied through simulation of effects on film thickness and adhesion quality. The thin film system and substrates used in this study are nickel-copper-silver deposited on 5/16 ANSI T5 ball bearings. The nickel-copper is sputter-deposited onto the substrates first, followed by evaporation-sputter deposition of pure silver. Rolling contact fatigue experiments using the coated ball-substrates were used to validate a model of a large ion plating system The balls were coated in the presence of pressure changes and arcing events commensurate with those simulated, and for different process recipes. The balls are then analyzed for film thickness and then life tested in a rolling contact fatigue tester similar to [12]Hoo. The rolling contact fatigue tests were carried out in high vacuum with a hertzian contact stress of 4.1GP at 130Hz rotational speed. The silver coating on the ball provides lubrication under high vacuum conditions, and the nickel-copper layer improves coating adhesion at the substrate surface.

Plasma sheath disturbances occur when arcing and subsequent pressure changes take place inside the process chamber. During these events a poorly tuned controller or a poorly designed process can introduce pressure and voltage spikes in the process gas by overshooting or under correcting for the event. In this paper a manufacturing ion plating process is simulated using a matrix sheath model and two types of process pressure control: PID and the Linear Quadratic Regulator (LQR). The PID model represents a control scheme for a typical ion plating process. The subsystem plant models have been validated individually and the process model presented in Figure 3 has been validated using regression analysis of a two factor central composite response surface.

DC PLASMA DESCRIPTION

A fluids based description of the plasma will be used in this paper. The fluid description treats the electrons and ions with a macroscopic fluids-based modeling approach. Group quantities and distributions are used to describe model variables such as ion velocity and energy. Quantum mechanical properties such as collision frequency and collision cross section were selected outside of the plasma model based on past glow discharge experiments and as reported in the literature.

PLASMA SHEATH

In this glow discharge process argon gas is ionized by inelastic electron-argon collisions which result in a change for both polarity and magnitude of the argon atom. The term "inelastic collision" means that one or both of the colliding particles have changed as a result of the collision. In contrast, the "elastic collision" between an argon ion and a neutral silver atom is strictly a momentum transfer action and neither particle has changed its material or electrical properties as a result of the collision. For more detail about ionization see [13]Lieberman.

The glow discharge can be separated into three primary regions: the anode sheath, a quasi-neutral region, and the cathode sheath. For a DC discharge process, the cathode is more biased than the anode due to charge conservation and the addition of the secondary electrons from the cathode surface that enter into the plasma sheath. Within the quasi neutral region, electrons carry most of the current and, since they do not have enough momentum to cross the sheath boundary, secondary electrons from the substrate surface satisfy charge continuity by poring into cathode sheath region from the cathode surface. The influx of substrate electrons collide with Argon atoms at the boundary between the quasi neutral region and the cathode sheath resulting in an intense ionization region, the so called pre-sheath. Within the cathode glow the electron density increases exponentially when moving inward from the cathode surface. The cathode sheath region represents a transient situation in which charge conservation must be satisfied in order to sustain the glow discharge.

The difference in potential across the plasma sheath directly influences the kinetic energy of the argon ions, which then creates the momentum transfer mechanism for imparting energy to the neutral and floating silver atoms near the substrate surface. Independence of the plasma regions allows one to model separately the sheaths and quasi neutral regions, with correct usage of boundary conditions, and removes the need to model the entire chamber in one simulation.

There are two types of DC sheath discharges, normal and abnormal. A normal discharge is observed in a spark or arcing event and it has the characteristic of a non uniform current density over the surface of the cathode. Abnormal discharges however exhibit uniform current density over the entire surface of the cathode, or substrate, and are well suited for ion plating processes. Because of their uniformity over the cathode surface, abnormal discharges can be sufficiently simulated using a 1D model. The ionization rate can be calculated from the current density at the sheath boundary and the voltage is a useful design parameter for system modeling. For example, an ionization rate of 4e15 ions/sec is needed to sustain an argon plasma at 60mTorr. This would require a current density of 0.3mA/cm^2 at 2000VDC. A system designer can select chamber and control hardware to deliver the needed current density and voltage.

The cathode sheath region of a DC discharge can be modeled using the Possion equation provided the assumption holds that the ion concentration is greater than the electron concentration within the sheath. The Possion equation has the form,

$$\nabla^2 \psi = \frac{e}{\varepsilon}\left(n_e - n_{Ar}\right) \tag{2}$$

where, n_e and n_{Ar} are the electron and argon ion concentrations. The sheath thickness is on order of a few Debye lengths, the pre-sheath is on order of a few mean free path lengths for ion-neutral collisions. Knowledge of the length of these regions enables robust insulator and high voltage isolation design inside the process chamber.

SPUTTER AND EVAPORATION-SPUTTER DEPOSITION

The process to be simulated has two mechanisms of deposition: (1) sputtering from a Ni/Cu target above the carousel and, (2) evaporation and sputtering of Ag from a crucible below the carousel. The evaporation-sputtering deposition begins when the source shutters are opened and silver atoms are allowed to migrate into the plasma sheath. Once inside the sheath argon ions impact the silver atoms onto the substrate surface. The shutters are situated over the silver crucible as shown in Figure 1. Deposition by sputtering begins as soon as the plasma is established and is dependant on process voltage and argon pressure. The first layers of Ni/Cu and subsequent Ag atoms are sputtered on and their properties are influenced by the plasma power, which is controlled by argon pressure and bias voltage. Atoms are liberated from the sputter target at a rate of,

$$R_{sp} = Y_{sp} V_{at} \left(J/q \right) \tag{3}$$

where Y_{sp}, V_{sp}, J, and q are the sputter yield, atomic volume, current density, and charge. For argon gas and silver the sputter yield is about 0.1 atoms per ion at 50eV. The controller design parameter for Equation (3) is the current density, J, which is influenced by the process voltage and argon gas pressure. During an arcing event the pressure in the chamber can rise by as much as 5mTorr which momentarily suppresses silver atom motion.

ELEMENTS OF THE ION PLATING PROCESS MODEL

Silver Evaporation Crucible Model

The evaporated silver available inside the chamber is modeled using a Hertz-Knudsen model. The Hertz-Knudsen model uses the temperature and vapor pressure of silver to calculate the amount of silver leaving the crucible as function of time,

$$\frac{dN}{dt} = \frac{C}{\sqrt{2\pi m/T}}\left(p^* - p(t)\right), \tag{4}$$

where the output dN/dt has units of atoms/cm^2/sec and with m, T, C, and p^* defined as the mass, temperature, view factor constant, and vapor pressure of the liquid silver.

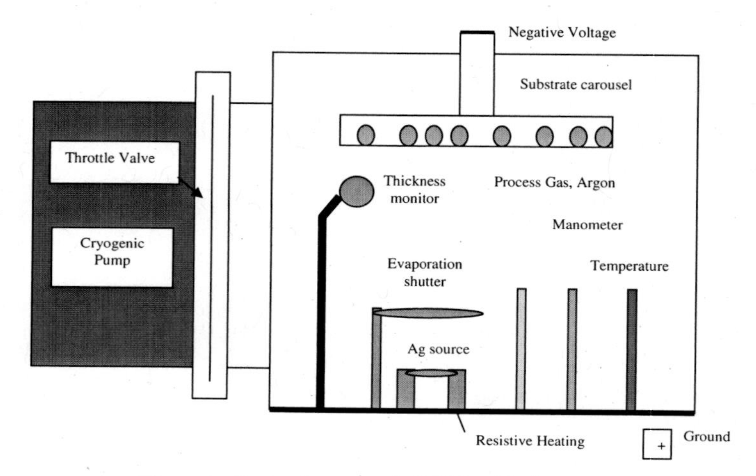

Figure 1. Basic elements of a DC Ion Plating process.

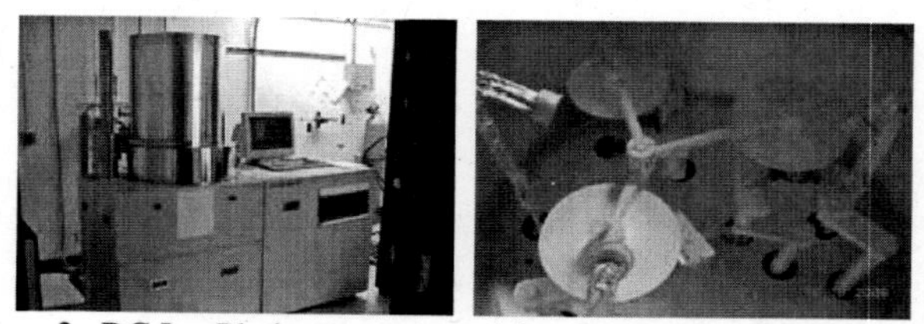

Figure 2. DC Ion Plating system used for simulation model validation.

Thickness Monitor Model

The plant model and controller for the thickness measurement system was taken from literature describing a typical monitoring system as found in [14]Sigma Instruments. The monitoring system plant is modeled as,

$$G(s)=\frac{K_p\exp(-Ls)}{T_1s+1}, \tag{5}$$

where K_p, L, and T_1 are the plant gain, system dead time, and time constant, respectively.

Typically, the thickness monitor is placed as close as practicable to the substrates inside the chamber and some calibration and tuning are required for accurate measurements. If the process voltage or pressure changes, then the thickness monitor needs to be re-tuned for the new process parameters. During arcing and pressure burst events, the thickness monitor measurement can be very inaccurate. A robust PID controller model is used and is described as,

$$C(s)=K_c\left(1+\frac{1}{T_is}+T_ds\right), \tag{6}$$

where K_c, T_i, and T_d are the PID controller gain, integral time, and derivative time, respectively. The model is presented using Simulink in Figure 3.

Process Pressure Models

The process pressure is maintained by two gas flow systems acting on the chamber simultaneously. A vacuum pump and conductance valve account for the gas removal portion of the pressure control system, and a gas introduction valve is used to introduce ultra high purity argon into the chamber. Typical volume flow rates for gas introduction are 60 to 200sccm, (Standard Cubic Centimeter per Minute). Most gas introduction systems are feed forward or feedback PD controlled subsystems. One common mass flow instrument uses a solenoid and field actuation coil to meter gas input, the details of which can be found in [15]O'Hanlon. The gas introduction system is presented below as a third order system that includes solenoid dynamics and a field coil actuator model,

$$\{\dot{x}\}=\begin{bmatrix}0 & 1 & 0\\ -k/m & -c/m & K_{mx}/m\\ 0 & 0 & -R/L\end{bmatrix}\begin{Bmatrix}x_1\\ x_2\\ x_3\end{Bmatrix}+\begin{bmatrix}0\\ 0\\ 1/L\end{bmatrix}V_x(t),\ Q_{in}=\begin{Bmatrix}K_f\\ 0\\ 0\end{Bmatrix}^T\{x\}. \tag{7}$$

The constants K_{mx} and K_f account for field current and gas flow calibration and are typically determined for a specific gas and flow range. The system output Q_{in} is in sccm and the state variables $\{x\}$ are the solenoid position, velocity, and field coil current.

The gas removal system is typically a feedback PID controlled scheme in which the valve conductance is modulated based on pressure monitor feedback information. A capacitance monometer is used to measure pressure electrically by monitoring the position of a diaphragm in response to pressure changes inside the chamber, [15]O'Hanlon. The conductance valve of the gas removal system consists of a DC motor drive with a gear reduction to open or close the conductance valve. The valve mechanism is modeled as a third order system that accounts for valve dynamics and the motor actuator. The output variable is the exiting mass flow, Q_{out} and is proportional to the valve-motor actuator position. The state space model is

$$\{\dot{z}\} = \begin{bmatrix} -R/L & 0 & 1/L \\ 0 & 0 & 1 \\ K_{mz}/J & 0 & -b/J \end{bmatrix} \begin{Bmatrix} z_1 \\ z_2 \\ z_3 \end{Bmatrix} + \begin{bmatrix} 1/L \\ 0 \\ 0 \end{bmatrix} V_z(t) \,,\; Q_{out} = \begin{Bmatrix} 0 \\ 1 \\ 0 \end{Bmatrix}^T \{z\} \tag{8}$$

where K_{mz} is the gate valve plant contstant.

The total pressure inside the process chamber is proportional to the difference of the incoming and exiting mass flows divided by the maximum pumping speed of the ultra high vacuum pump, [15]O'Hanlon. In this model the pressure is described as,

$$P = \frac{Q_{in} - Q_{out}}{S} \tag{9}$$

where, S is the throughput of the pumping system at process pressure. Performance curves and throughput at different base pressures are used to size the pumping systems with respect to flow rate capability at the desired base pressure.

Cathode DC Sheath Model

At potentials in the kV range the cathode fall region is abnormal and the energy distribution of electron flux, EDEF, is uniform across all cathode surfaces and a 1D sheath model is sufficient to calculate ion current density at the substrate surface. A matrix sheath model from [13]Lieberman is used and is based on a solution to Equation (2). The current density at the sheath and cathode interface is

$$J = en_i\bar{u}_i \tag{10}$$

where, e and $\bar{u}_i$ are electron charge and mean ion velocity, respectively. The density n_i is the ion density in the sheath, similar to that used in (2). The group velocity is calculated as,

$$\bar{u}_i = \left(\frac{eV_0\pi\lambda_i}{m_i s} \right)^{1/2} \tag{11}$$

where V_0, λ_i, m_i, and s are the process DC voltage, ion mean free path, ion mass, and sheath thickness. The ion mean free path is affected by the total gas pressure in the chamber and defined as,

$$\lambda_i = 1/n_g\sigma_i, \tag{12}$$

where n_g is the gas density and σ_i is the ion collision cross section. The parameter σ_i accounts for the quantum mechanics of the ionization process and can be found tabulated in the literature. The sheath thickness is calculated as,

$$s = \sqrt{\frac{2\varepsilon_0 V_0}{e n_i}} \tag{13}$$

where ε_0 is the permittivity of free space constant. The electric field across the sheath is calculated as,

$$E = \left[\frac{3 e n_i \bar{u}_i}{2\varepsilon_0}\right]^{2/3} \frac{x^{2/3}}{\left[2e\lambda_i / \pi m_i\right]^{1/3}}. \tag{14}$$

Equations (2) and (10) through (14) are used in the ion plating plant model in Figure 3 to calculate ion current density and kinetic energy at the sheath and substrate boundary. A constant ion density taken from the pre-sheath calculation is used in Equation (13) to reduce numerical complexity.

ION PLATING MODEL THICKNESS TEST

Regression analyses using central composite design and response surface analysis was used to validate the process model shown in Figure 3. The variable factors were argon pressure and DC bias voltage. The deposition rate and thickness monitor parameters for all experiments in the Design of Experiments (DOE) were the same. All DOE runs were stopped when the thickness monitor measured 110nm, which resulted in a similar overall process time of about 20 minutes. A baseline process map is presented in Figure 4. The process voltage and pressure were adjusted as per the values presented in Table 1. For example, experiment 3 was run at 17.5mTorr and the voltage profile in Figure 4 was shifted down 1.0kV. Experiment 12 was run at 15.0mTorr with the voltage profile shifted up 1.0kV. Final thickness was determined by weight comparison of pre and post coated balls. Sixty balls were used for each DOE run and an average thickness of silver based on weight was assumed for each ball. All balls for all the tests came from the same lot of ANSI T5 and were preprocessed similarly.

All experiments were run in power control using the system presented in Figure 2 and this data was used in the regression model presented in Figure 5. Considering Figure 5, at lower voltages the thickness monitor under predicts true thickness and for higher voltages the thickness monitor over predicts true thickness. This is understood by considering the effect of process pressure on the motion of evaporated silver from the crucible. There is less deposition when the process pressure or the voltage is increased due to the reduced thickness of the cathode sheath and the increased sputtering action at higher process voltage.

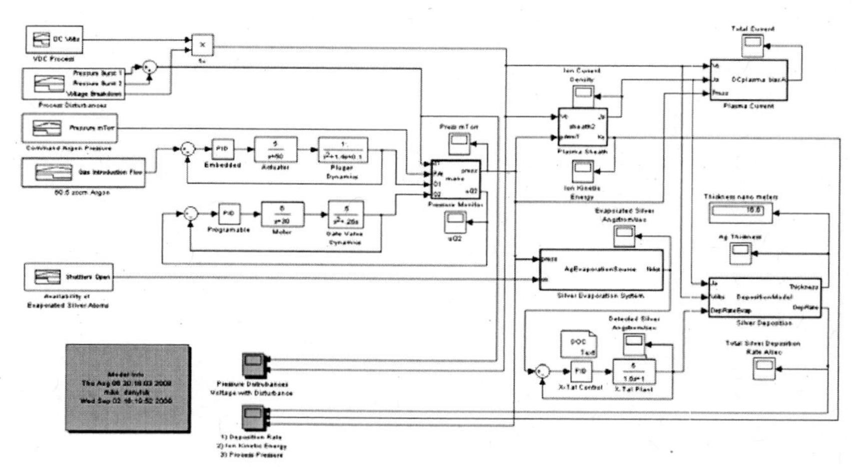

Figure 3. Block diagram of a typical ion plating system using PID control.

Interestingly, for voltages above 2.8kV the effects on thickness measurement are dominated by the process voltage. This can be explained by the increased sputter energy above 2.8kV that effectively removes as much or more silver than is deposited onto the substrates.

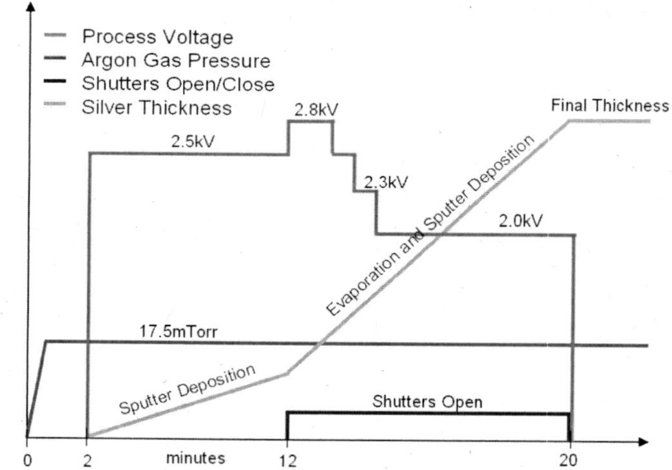

Figure 4. Process map of baseline ion plating process for DOE tests.

Table 1. DOE description and thickness results.

RunOrder	kVolts	mTorr	Thickness nm
1	3.5	18.5	89.70
2	2.5	17.5	101.80
3	1.5	17.5	108.90
4	2.5	18.5	Not tested
5	1.5	18.5	104.90
6	2.5	17.5	101.90
7	2.5	17.5	102.30
8	1.5	15.0	137.60
9	3.5	17.5	78.70
10	2.5	17.5	102.10
11	2.5	15.0	Not tested
12	3.5	15.0	66.50
13	2.5	17.5	101.60

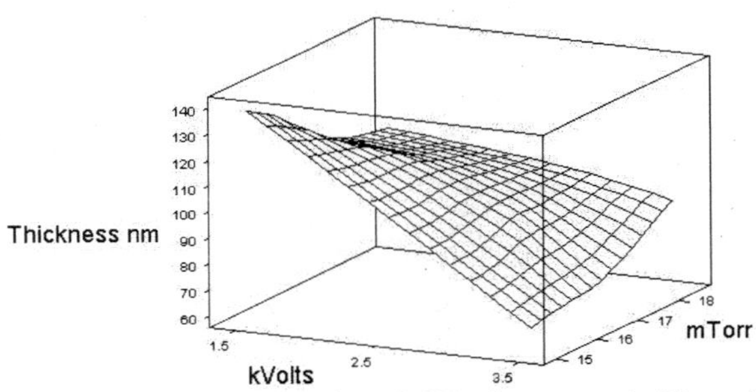

Figure 5. Response surface from DOE using system in Figure 2.

SEM RESULTS OF EXTREME DOE TESTS

Table 2 contains SEM elemental results of four DOE runs that represent the extreme values of the DOE variables. The Ni and Cu present in the results are from a Monel 400 sputter plate that was mounted above the substrate carousel. The Ni and Cu act as tracers to help determine the balance between sputter source and deposition as the process variables are changed. For example, Run 2 has trace amounts of Ni and Cu, suggesting that the sputter plate contributed to deposition on the balls. Run 8 however, which used a lower power process, has less Ni and Cu than Run 2, suggesting that at 1.5kV and 15mTorr the sputter plate is not a significant source of deposition material. In contrast, Run 12 has a small amount of Ni and Cu present as well as less silver and was run at 3.5kV and 15mTorr suggesting that the sputtering was too harsh even at the same peak power as Run 2, and resulted in an overall reduced thickness. [17]Kolev et al. calculate an increase in ion energy as pressure decreases and also confirm that there is less redistribution of sputtered material with increasing gas pressure for the ranges of 4 to 100 mTorr. The presence of Ar in Run 2 was confirmed during RCF testing in vacuum using RGA. The Ar gas gets trapped inside the coating layers during deposition. Figure 6 is an SEM image of one ball from DOE Run 2.

Table 2. SEM results from one ball of four DOE results and one uncoated ANSI T5 ball.

Element	Base T5 no coating	DOE Run 2 101.8nm Coating Peak Power 1000W 17.5mTorr 2.5kV	DOE Run 3 108.9nm Coating Peak Power 338W 17.5mTorr 1.5kV	DOE Run 12 66.5nm Coating Peak Power 1000W 15.0mTorr 3.5kV	DOE Run 8 137.6nm Coating Peak Power 180W 15.0mTorr 1.5kV
Ar	0.00	0.82	0.00	0.00	0.00
Ni	0.00	10.89	2.70	0.56	0.45
Cu	0.00	3.33	0.84	0.30	0.12
Ag	0.00	24.13	39.45	19.54	31.22
Fe	65.82	45.12	39.83	63.68	53.15
Cr	4.15	3.72	3.26	4.03	3.69
V	1.25	1.88	1.08	1.13	1.04
W	18.00	10.11	12.84	10.76	10.33
C	0.78	0.00	0.00	0.00	0.00
Co	10.00	0.00	0.00	0.00	0.00

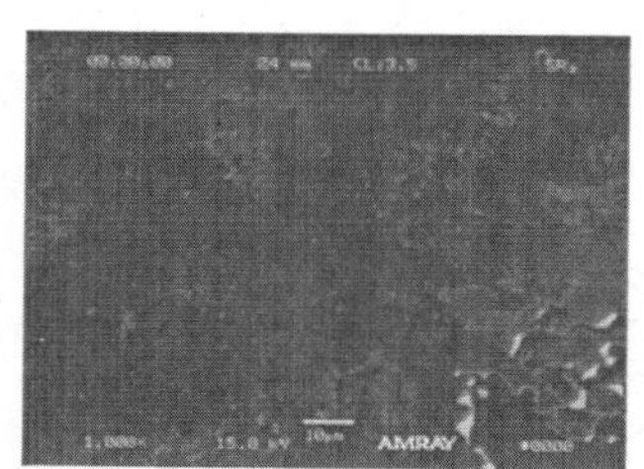

Figure 6. SEM image of baseline process Run 2 of Table 1.

ROLLING CONTACT FATIGUE TESTING OF EXTREME DOE COATED BALLS

A rolling contact fatigue (RCF) test similar to [12]Hoo is used to quantify thin film adhesion quality and life. The test platform has been modified for solid lubrication systems and uses 6 coated balls per test. The tests were conducted in UHV using a fixed load and speed for all tests. The contact stress is calculated as 4.1GPa with rotation speed of 130Hz.

The solid lubrication UHV-RCF test platform has been validated for repeatability using 5 tests on a separate set of silver coated balls from the same lot and process history. A fatigue life curve of repeatability tests is presented in Figure 7. A screen shot of the data monitoring system results and a post test photo of one ball is presented in the inset of Figure 6 as well.

Two categories of failure modes were observed when testing the DOE coatings in Tables 2 and 3. The first observed failure mode was silver depletion followed by substrate spalling after about 32 hours. This mode is similar to failures observed in the repeatability tests in Figure 7. The second observed failure mode was spalling of the coating within 6-8 hours. Four DOE recipes have been UHV-RCF tested so far, Runs 2, 3, 8, and 12, as these represent the extremes of the experimental design space with respect to process pressure and voltage.

The second failure mode was unique in that the balls did not precess during the test. Effectively the balls ran on one track about the O.D. of the ball instead of precessing and running over the entire surface of the ball. This suggests a change in lubrication properties of the silver coating, which was brought about by process pressure and voltage changes during the coating deposition process.

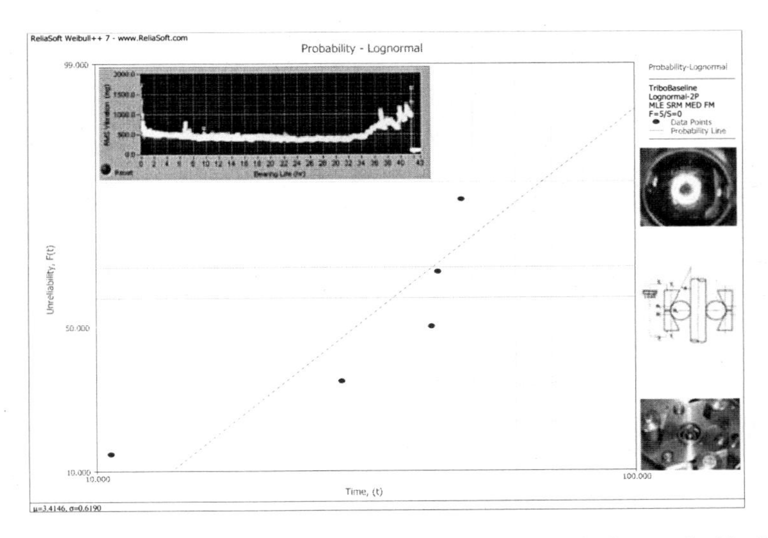

Figure 7. Repeatability tests of UHV-RCF tester using ANSI T5 balls coated with silver.

Interestingly, spalling failure was observed in [9]Liston for TiN coatings on M50 steel rods but with a different outcome due to the differences between solid and liquid lubrication testing platforms. The main difference is that in the solid lubrication system the spall material remains in the race track for the life of the test whereas in the oil drip system, the oil washes away spall material from race track allowing the test to continue.

Table 3. UHV-RCF test results in hours.

RCF Test	DOE Run 2 101.8nm Coating Peak Power 1000W 17.5mTorr 2.5kV	DOE Run 3 108.9nm Coating Peak Power 338W 17.5mTorr 1.5kV	DOE Run 12 66.5nm Coating Peak Power 1000W 15.0mTorr 3.5kV	DOE Run 8 137.6nm Coating Peak Power 180W 15.0mTorr 1.5kV
1	8.1	8.2	3.3	8.0
2	6.5	32.0	4.5	7.1
3	6.9	8.5	Not tested	30.1
4	6.6	10.1	Not tested	5.5
5	6.4	7.5	Not tested	11.9

The longest test time results came from DOE Runs 3 and 8 with 32 and 30.1 hours to failure. Balls from DOE Run 2 failed consistently around 6.5 hours based on 5 tests. The shortest test times came from Run 12. These data suggest that increased process voltage decreases RCF life. More interesting was the nature of the failure of the coatings from Runs 2 and 12. These tests showed little ball precession suggesting different lubrication properties. The process used in DOE Runs 3 and 8 showed the most promise in that the coating was depleted evenly over time, and these tests were the longest.

SIMULATION RESULTS

Arc disturbances that resulted in pressure increases and plasma break down were accounted for in the ion plating model simulation. Based on the DOE and UHV-RCF results the controller must mitigate disturbance effects during deposition. The same disturbance events were used on the plant model in Figure 3 but for two different controller schemes, namely PID and LQR.

Output Data of System with Real Disturbances

Output data from the system presented in Figure 2 is shown in Figure 8 below. The process is slightly different from the DOE process presented in Figure 4 and is used here for comparison only with a historical process carried out on the ion plating system in Figure 2. The decrease in bias current over time in Figure 8 is due to the removal of oxides on the cathode surfaces. The oxides have a larger secondary electron yield than does the underlying cathode surface, which results in a larger bias current at the beginning of the process. As the process proceeds, the argon ions remove the oxides from the surface and the bias current settles to levels associated with the secondary electron yield of the substrate and the sputter target base material. The process pressure data in Figure 8 contains arcing events and their influence on the bias current and pressure during a process. The model presented in Figure 3 does not account for oxides on the cathode and substrate surfaces.

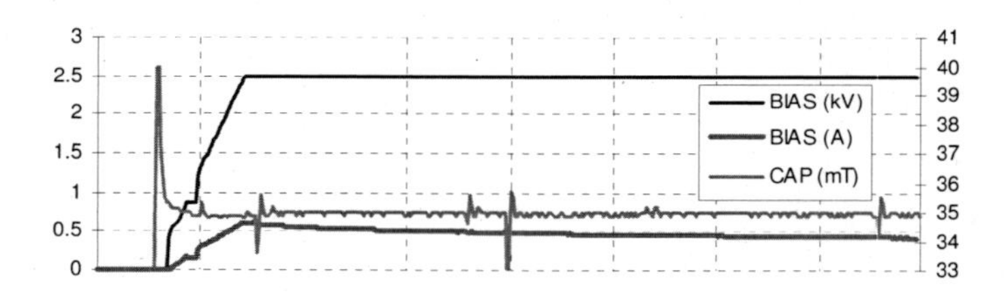

Figure 8. Output data from the system in Figure 2.

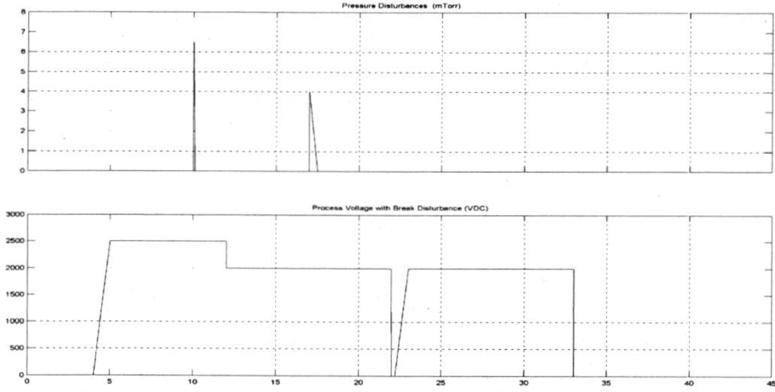

Figure 9. Disturbance inputs and voltage profile to system in Figure 3.

Simulation results from PID and LQR model control.

An LQR control scheme following [16]Burl was implemented on the gas removal system presented in Equation (8) and used in the model presented in Figure 3. The input disturbances include two pressure bursts and one high voltage breakdown event. The disturbances are presented in Figure 9. The output response of the system for the two controller types is presented in Figure 10 for comparison.

The LQR control scheme allows the system designer to optimize performance of the valve by assigning penalties to the control law variables. For example, in the present study there is no way to completely mitigate pressure bursts inside the process chamber due to their rapid and random nature. However, system behavior after the burst subsides, usually within 1 or 2 seconds, can be controlled such that the control system need not further react to the event. The difficulty lies with the high inertia properties associated with vacuum valve hardware. The LQR controller used in this study was optimized for minimum overshoot. In contrast, the PID controller tries to respond to the pressure disturbance and subsequently overshoots, which affects deposition rate and ion kinetic energy as shown in Figure 10. [17]Kolev et al. simulate a decrease in ion kinetic energy with increase in process pressure as well, due to the increased collisions with argon atoms associated with the pressure increase. No attempt was made to correct for voltage disturbance it was presented here to show its influence on the process output.

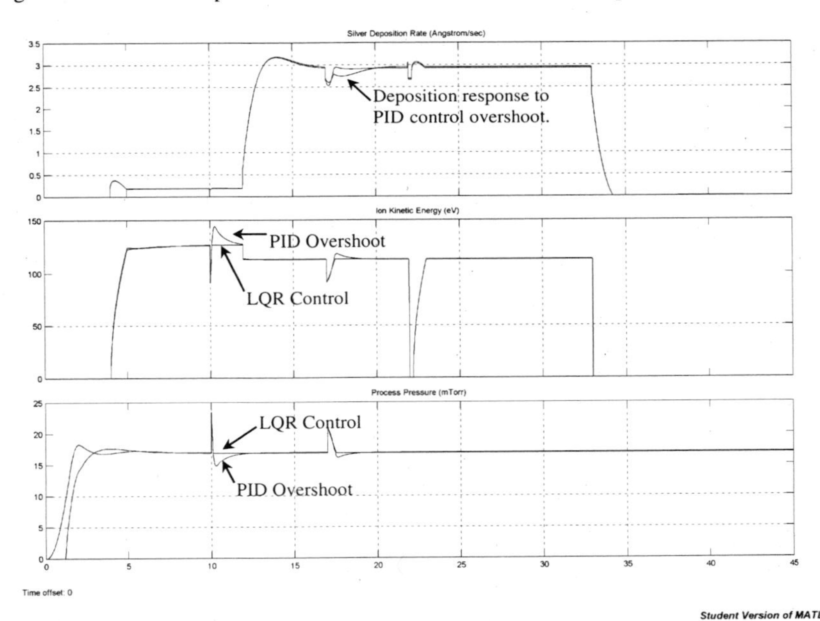

Figure 10. Output from PID and LQR Control the conductance valve in the system of Figure 3.

CONCLUSIONS

High voltage events can remove deposited material from substrates and change the elemental composition of the coating. The structure and composition of a PVD coating can be altered based on sputtering power as reported in the literature. For the process recipes studied in this paper, the combination of 17.5 and 15mTorr argon pressure and nominal voltage of 1.5kV yielded the longest UHV-RCF test times, 32 and 30.1 hours. The combination of 17.5mTorr and 2.5kV yielded an UHV-RCF test life of around 6.5hrs and consistently exhibited a non-precession type failure mode for all tests. These data suggest that higher process voltages during deposition have a detrimental effect on coating adhesion and lubrication properties. A higher density film is deposited at 2.5 and 3.5kV, but it is unclear whether the increased density has a detrimental effect on UHV-RCF life or if changes in composition reduce UHV-RCF life. In contrast, at 1.5kV process voltage, a less dense film was deposited and this had a positive influence on UHV-RCF life. Further study is needed to understand in detail the effects on lubricity from process voltage and pressure.

A rolling contact fatigue test platform operated in ultra high vacuum was introduced in this paper. The UHV-RCF test provides a consistent and repeatable testing platform for solid lubrication systems. The next step is to test silver coated M50 balls and rods, similar to the industry standard data for liquid lubrication systems, and compare results with respect to model strategy.

Regression analysis of 9 DOE runs in which pressure and voltage were varied revealed significant errors in coating monitor thickness results. At higher voltages the thickness monitor over predicted coating thickness and at lower voltages the monitor under predicted thickness. For voltages over 2.8kV, sputter removal surpasses evaporation deposition and the coating gets thinner, even at lower pressures for which total power is maintained as compared to other DOE runs.

Ion plating processes require multiple systems to function simultaneously during pressure burst and arcing events. Since the subsystem PID controllers are tuned around one specific operating point, errors can occur within the process if chamber conditions change beyond the tuning space of the controllers. Plant disturbances such as pressure bursts due to arcing cannot be mitigated a priori and an optimization strategy like the LQR control could be used to reduce disturbance response time.

REFERENCES

[1]D. Mattox, Handbook of Physical Vapor Deposition Processing, Noyes Publications, Westwood New Jersey (1998).
[2]C. Zechner, C. Moroz, Simulation of doping profile formation: Historical evolution and present strengths and weaknesses, *J. Vac. Sci. Technology B*. Vol. 26, Issue 1, pp. 273-280, (2008).
[3]W. Moller, W. and D. Guttler, Modeling of plasma-target interaction during reactive magnetron sputtering of TiN, *Journal of Applied Physics* Vol. 102, Article 094501, pp. 1-11.(2007).
[4]Q. Qiu, Q. Li, J. Su, Y. Jiuo and J. Finley, Simulation to Predict Target Erosion of Planar DC Magnetron, *Plasma Science and Technology*, Vol 10 No 5 (2008).
[5]J. Moyne, E. DelCastillo and M. Hurwitz, Editors, Run to run Control in Semiconductor manufacturing, CRC Press (2001).
[6]W. Chan and E. Chason, Stress evolution and defect diffusion in Cu during low energy ion irradiation: Experiments and Modeling, *J. Vac. Sci. Technol. A* 26(1) (2008).
[7]M. Kadera, S. Ito, M. Hasunuma, and S. Kakinuma, 2007, Nanometer-Scale Stress Field Evaluation of Cu/ILD Structure by Cathodoluminescence Spectroscopy, Stress induced phenomena in metallization, Kyoto Japan, AIP conference proceedings (2007).
[8]J. Bao, H. Shi, J. Liu, H. Huang, P. Ho, M. Goodner, and G. Kloster, Mechanistic Study of Plasma Damage of Low k Dielectric Surfaces, Stress induced phenomena in metallization, Kyoto Japan, AIP conference proceedings (2007).
[9]M. Liston, Rolling Contact Fatigue Properties of TiN/NBN Coatings on M-50 Steel, J. Hoo and W. Green, editors, ASTM STP1327 (1998).
[10]C. Liu, and Y. Choi, Rolling contact fatigue life model incorporating residual stress scatter, *International Journal of Mechanical Sciences*, vol 50, issue 12 (2008).
[11]V. Felmetsger, P. Laptev, and S. Tanner, Innovative technique for tailoring intrinsic stress in reactively sputtered piezoelectric aluminum nitride films, *J. Vac. Sci. Technol. A* 27(3), May/Jun (2009).
[12]J. Hoo, A Ball-Rod Rolling Contact Fatigue Tester, ASTM STP 771 ASTM pp. 107-124 (1982).
[13]M. Lieberman, and A. Lichtenberg, Principles of Plasma Discharges and Materials Processing, 2nd Ed., John Wiley & Sons, Inc (1994).
[14]Sigma Instruments, SQC-300 Thin Film Deposition Controllers, (2007).
[15]J. O'Hanlon, User's Guide to Vacuum Technology, 2nd Ed., John Wiley & Sons, Inc (1989).
[16]J. Burl, Linear Optimal Control, Addison-Wesley. (1999).
[17]I. Kolev and A. Bogaerts, Numerical study of the sputtering in a dc magnetron, *J. Vac. Sci. Technol. A* 27(1), Jan/Feb (2009).
[18]O. Malyshev, R. Valizadeh, J. Colligon, A. Hannah, K. Middleman, S. Patel, and V. Vishnyakov, Influence of deposition pressure and pulsed dc sputtering on pumping properties of Ti-Zr-V nonevaporable getter films, *J. Vac. Sci. Technol. A* 27(3), May/Jun (2009).
[19]N. A. Morley, S.L. Yeh, S. Rigby, A. Javed, and M.R.J. Gibbs, Deposition of a cosputter-evaporation chamber for Fe-Ga films, *J. Vac. Sci. Technol. A* 26(4), Jul/Aug (2008).

Author Index